Mik

GEOGRAPHERS ABROAD

Essays on the Problems and Prospects of Research in Foreign Areas

---

Marvin W. Mikesell, *Editor*

H. C. Brookfield
Paul Ward English
C. A. Fisher
Chauncy D. Harris
David Lowenthal
Rhoads Murphey
Norton Ginsburg
James J. Parsons
Edward W. Soja
David E. Sopher

THE UNIVERSITY OF CHICAGO
DEPARTMENT OF GEOGRAPHY
RESEARCH PAPER NO. 152

1973

Published 1973 by The Department of Geography
The University of Chicago, Chicago, Illinois

Library of Congress Catalog Card Number: 73-87829

Research Papers are available from:
The University of Chicago
Department of Geography
5828 S. University Avenue
Chicago, Illinois 60637
Price: $5.00 list; $4.00 series subscription

PREFACE

This volume is the product of a conference held at the University of Chicago in November of 1972 under the sponsorship of the Harris Foundation. The purpose of this foundation, established through the generosity of the heirs of Norman Wait Harris and Emma Gale Harris in 1923, has been to promote "a better understanding on the part of American citizens of the other peoples of the world." In keeping with this objective, the Foundation offered support for a conference designed to provide an opportunity for discussion of some of the problems associated with geographic research in foreign areas. The rationale implicit in this design was influenced by several considerations.

In the first place, it is not possible to point to any general review of the contribution geographers have made or might make to the vast enterprise of foreign area studies. To be sure, geographers devoted to particular foreign areas are usually well aware of the accomplishments of their colleagues, and this awareness is enhanced by the meetings and publications of such organizations as the African Studies Association, the Association for Asian Studies, the Middle East Studies Association, and so on. Nevertheless, the question often posed but seldom answered of how work undertaken in one realm of culture can be related to work conducted elsewhere leaves most area specialists with feelings of confusion and inadequacy. Hence it was hoped that a collection of essays designed to demonstrate the character of geographic research in several foreign areas might provide a foundation for comparison.

The rationale of the conference was also influenced by the fact that geographers who hope to work abroad must now accept a number of discouraging realities. Once among the most prosperous of academic enterprises, foreign area studies have suffered from a curtailment of support in recent years. Moreover, many countries are now imposing restrictions that limit the freedom of movement and access to research materials that are essential requirements of geographic scholarship. These developments do not, of course, mean that research in foreign areas is no longer possible; yet it would be unrealistic to expect that research can be conducted on the vast scale of the 1950's and 1960's.

Hence it seems reasonable to suppose that greater effort should be made to identify the additive and cumulative features of foreign area research in order that more effective use might be made of surviving opportunities. Needless to say, it would not be wise in any scholarly enterprise to depreciate the virtue or minimize the power of "rugged individualism." Yet one can also proclaim the importance of and indeed the necessity for an assessment of neglected, completed, and nearly completed tasks. The essays in this volume represent an attempt to offer or at least initiate an assessment of this character.

On the other hand, the fact that foreign area research must be viewed in the context of a recent transition from "boom" to "bust" should not be allowed to obscure the historical depth and vitality of this commitment. Among German geographers, especially, the tradition that university careers should be built on a foundation of prolonged field study in foreign areas is strikingly evident in the life and work of such academic pioneers as Ferdinand von Richthofen, Friedrich Ratzel, and Karl Sapper. To be sure, encouragement of foreign area studies (*Auslandskunde*) was balanced in that academic tradition by the respect accorded to scholars who acquired a deep understanding of the homeland (*Heimatkunde*). Still it is difficult to think of the German geographic tradition without also thinking of the quest for understanding of foreign areas that was so brilliantly initiated by Alexander von Humboldt.

In France, similarly, the academic program established by Paul Vidal de la Blache included systematic examination of major world areas (*géographie universelle*) as well as more detailed studies of the countless *pays* of the homeland. It is hardly surprising, therefore, that the geographical traditions of Britain and the United States should also have included a commitment to seek understanding among alien peoples and in exotic environments. Moreover, British geographers, like their French colleagues, were able to acquire overseas experience as a result of academic appointments in Baghdad, Khartoum, Hong Kong, Rangoon, Singapore, and other administrative and intellectual centers of their Empire.

It goes without saying that such opportunities were not open to citizens of the United States until fairly recently. Yet grants from universities and foundations, personal fortunes, and missionary affiliations permitted some Americans to work abroad even during the interval between the first and second world wars, when national interest in foreign areas was minimal. For example, the list of dissertations published in the *Annals of the Association of American Geographers* in 1935 indicates that 35 of a total of 136 studies completed prior to that time were devoted to areas outside the United States and Canada. In other words, the area and language centers established in the United States after the second world war did not initiate a commitment for American geographers; they merely encouraged a larger group of scholars to engage in studies that had always been part of the heritage of geography.

Still it cannot be denied that the costly infrastructure built up in the United States during the 1950's and 1960's resulted in a substantial increase in the number of geographers who were able to work abroad. For example, approximately 400 of the 1582 titles listed in Clyde Browning's Bibliography of Dissertations in Geography, 1901-1969 (Department of Geography, University of North Carolina, 1970) refer to areas outside the United States and Canada and about half of these efforts seem to have entailed field study or at least residence abroad.

The fact that only one of the programs that made this commitment possible, the Foreign Field Training Program of NAS-NRC (1955-1969), was exclusively for geographers means that standards in regard to what constitutes adequate preparation for work abroad were established for the most part by scholars trained in disciplines other than geography. The Foreign Area Fellowship Program initiated in the early 1950's by the Ford Foundation was particularly emphatic in encouraging the idea that the training requirements of foreign area specialists should include a quadruple program. It was thought that they should have a working knowledge of a particular foreign language, be liberally educated in the history of a particular civilization, and also have an effective command of both a major and complementary discipline. The difficulties entailed in such programs were forbidding; yet the prospect of generous rewards, even to the extent of completely subsidized graduate student careers, precluded serious criticism. The fact that such rewards are less likely today means that professors devoted to the cause of foreign area specialization will have to reconsider a number of assumptions made with this program in mind.

Finally, any review of the many issues having to do with the character of geographic research in foreign areas must include some consideration of the shifts of emphasis that have occurred within the profession in recent years. The presumed contrast of "old" and "new" geography has been debated at such length that even the most carefully reasoned arguments seem to offer little more than caricatures and clichés. For the subject at hand the most important question has to do with whether methods, theories, and models developed to solve research problems in the Occident can be applied with equal or at least acceptable results in non-Western areas. To date the response to this question has been equivocal--"yes maybe" and "yes" and "no." Moreover, underlying this question and the partial answers offered in response to it, are philosophical issues of pervasive importance. How many exceptions are required to disprove a rule? How much proof is needed to confirm an experiment? How may demonstrations of correlation be translated into demonstrations of causation?

The fact that these issues, which are the essence of science, may inspire highly personal responses, and hence become articles of faith, has inhibited communication among geographers and doubtless will continue to do so. Yet dialogue even among "believers" can be improved if it is realized that creative tension between the particular and the general is an inherent feature of all scholarship. The author of a specific case study should expect to be challenged to expand his findings; and the author of a generalization should expect to be challenged by exceptions.

The conference held at the University of Chicago was designed to provide an occasion for critical review of these and other issues of major importance to geographers who have worked or hope to work abroad. The participants were asked to respond to the challenge of the conference in any way that seemed appropriate. With the exception of Professor Fisher, who was not able to come to Chicago, each of the contributors distributed copies of his paper prior to the conference, which, consequently, was devoted entirely to discussion.

A summary of that discussion, which appears as the final part of this volume, was prepared by the editor from tape recordings. While this account does not capture every nuance of the conference and presents, in paraphrase, only about a quarter of its words, it does offer a full inventory of the topics considered and reproduces faithfully the sequence of the discussion. That the conference was relatively unstructured is self-evident, for although each author was asked to summarize his paper, the subsequent discussions were not thus confined, for questions were posed and comments offered according to a spontaneous order. For example, although the issue of the importance of language training was first raised in the context of research in Latin America, it was brought up on several other occasions and hence became a leitmotiv of the conference. Another issue raised repeatedly was the fascinating question of the "personality" of geographic research in the several areas, and whether this might reflect culture and environment or accidents of individual or institutional leadership. The validity of the area studies framework was both warmly endorsed and severely criticized during the conference; yet even those who were critical of this framework were enthusiastic about what had been done or might be done in "their areas."

The conference concluded--not surprisingly--with an endorsement of cross-cultural comparison as a desirable and attainable objective for foreign area studies. Recognition of the importance of this objective seems to have been at or close to the surface of the conference at all times. And comparison was attempted on several occasions. The conference also produced some sobering comments on the difficulties of comparison.

The members of the Department of Geography at the University of Chicago are grateful to the Harris Foundation and to Bert F. Hoselitz, Chairman of its faculty committee, for enabling them to welcome a distinguished group of colleagues. If the gentlemen who gathered here in November of 1972 are at all representative of geographers who have worked abroad, this particular species of academic man is unusually articulate and congenial.

Marvin W. Mikesell
August, 1973

TABLE OF CONTENTS

CHAPTER 1

GEOGRAPHIC STUDIES OF THE SOVIET UNION: SOME REFLECTIONS

Chauncy D. Harris
University of Chicago

Rather than present a scrupulously balanced, objective, and comprehensive review of the contributions of geographers to area studies of the Soviet Union, I shall offer some personal criticisms of geographic interpretations of the Soviet Union, which, in my mind, illustrate five pitfalls or dangers in the study of a foreign area or culture. This will be followed by a statement of some needs in future geographic work on the Soviet Union and finally by a suggestion on the comparative role of scholars from within the area and of those from outside.

The Danger of Inferring from One's Own Conceptions Characteristics of Another Culture

Geographers and historians of the West have explained the expansion of the Russian Empire as an urge to the sea without bothering to examine whether such a striving was expressed by the Russians themselves.[1]

George B. Cressey, an early American student of the Soviet Union who travelled extensively in the Soviet Union but who was not able to read the Russian literature, stated quite flatly: "The history of Russia may be written in terms of its search for ocean ports. The Russian Bear will not be content until it finds warm water . . . Pressure against Persia has continued for more than a century, partly with the hope of securing a port on the Persian Gulf . . . Intrigues in Afghanistan, Tibet, Mongolia, Manchuria, and elsewhere in China are parts of the same inescapable programs. Late in the nineteenth century the Trans-

[1] John A. Morrison, "Russia and Warm Water, a Fallacious Generalization and Its Consequences," United States Naval Institute Proceedings, Vol. 78, No. 11 (November, 1952), pp. 1169-1179.

Siberian Railway was built as another attempt to get at the open sea . . . Soviet foreign policy may be expected to have freer access to the ocean as one of its permanent ambitions."[2]

Isaiah Bowman, a leading American student of political geography, in his magisterial volume The New World[3] suggested that "In Mongolia and Tibet, in Persia and Afghanistan, in Caucasia and at Constantinople, the Russian has been pressing forward for three hundred years, and no system of government can stand that denies him proper outlets. His slogan has been a 'warm water port.' That explains his reaching out in the Far East to Vladivostok . . . it explains his effort to reach the head of the Persian Gulf . . . ; it explains the struggle with Turkey and the West-European powers for Constantinople."

The German geographer Erich Thiel, repeated the same theme: "In no land on the earth is the natural striving toward the sea so clearly expressed in history as in Russia."[4]

Even the historian Robert J. Kerner mentions the "Russian urge to the sea--a centuries-old national longing--because its independence and security depended upon access to the seas."[5]

The consensus among Western authorities seems overwhelming. But I submit that solid evidence from the literature of the Russian Empire or the Soviet Union is lacking. Except for the very special case of Peter the Great seeking a "window to the West" in St. Petersburg on the Gulf of Finland, the Russian literature is remarkably devoid of concern for access to the ocean, for warm-water ports, or especially for expansion outward in order to reach the sea. Russia has been a continental power concerned mainly with internal prob-

---

[2]George B. Cressey, The Basis of Soviet Strength (New York: McGraw Hill, 1945), p. 242.

[3]Isaiah Bowman, The New World: Problems in Political Geography (Yonkers-on-Hudson, New York: World Book Company, 1921), p. 398. (This quotation was eliminated from the fourth edition, published in 1928.)

[4]Erich Thiel, Sowjet Fernost: eine landes- und wirtschaftskundliche Übersicht (München: Isar Verlag, 1953), p. 12.

[5]Robert J. Kerner, "The Soviet Union as a Sea Power," in New Compass of the World, ed. by Hans W. Weigert and Viljhalmur Stefanssen (New York: Macmillan, 1949), p. 104. This theme finds expression in the title of The Urge to the Sea: the Course of Russian History: the role of rivers, portages, ostrogs, monasteries, and furs (Berkeley, California: University of California Press, 1942), 212 pp., but the subtitle reveals the real content of the book, which has nothing to do, really, with any urge to the sea.

lems and with development of resources within its own territory. Expansion has been for more agricultural land, for furs, for security against nomadic raiders or border states, and for protection of Christian peoples against Moslem neighbors, rather than for territory that extends to the open ocean. Certainly the expansion across Siberia had nothing to do with any urge to the sea; only centuries later, long after Russia had a long coastline on the Pacific Ocean and the Sea of Okhotsk, did concern arise for a base for the Far Eastern fleet in Vladivostok. Sea-conscious leaders attributed their ideas to a land-based people. Winston Churchill, First Lord of the Admiralty, and Franklin Delano Roosevelt, Assistant Secretary of the Navy, both in the First World War, saw the world from the sea and during World War II doubtless incorrectly inferred from their perceptions of the key interests of Joseph Stalin, a Georgian mountaineer concerned mainly with the internal development of the Soviet Union.

The Deceptive Insights of Grand Generalizations

As early as 1904 in a lecture on "The Geographical Pivot of History" before the Royal Geographical Society, Halford J. Mackinder developed his concept of the "World island" consisting of Europe, Asia, and Africa, within which there was a core of "heartland" in the northern interior of Eurasia. In his interpretation of world politics as a struggle between sea power and land power, this heartland was so remote and inaccessible from the ocean and hence from naval power as to be able to withstand indefinitely a siege from a sea power. He further wrote "Who rules the Heartland commands the World-Island. Who rules the World-Island commands the World."[6] Without in any way wishing to underestimate the importance of space and depth in military affairs, as both were important in the survival of the Soviet Union in World War II, I wish to suggest that this grand generalization does not provide a useful insight into the areal patterns of political and military power. His heartland consists of vast deserts of Middle Asia and trackless forest of Siberia. The most remote area from sea power is northern Siberia, devoid of railroads, roads, cities, or major industries that might sustain or provision a large military force. The real centers of economic, political, and military power in Eurasia lie not here but in more fertile, more densely populated, more highly developed industrial lands to west,

---

[6]Halford J. Mackinder, Democratic Ideals and Reality: A Study in the Politics of Reconstruction (New York: Holt, 1942), p. 150 (first published in 1919).

south, and east in the European Soviet Union, in Central and Western Europe, in India in Southern Asia, and in China and Japan in Eastern Asia. It is people and production not mere space that is the basis of power. The economic heart of the Soviet Union, which occupies the interior core of Eurasia, lies primarily not in areas merely remote from the sea but rather in developed areas with agriculture, mining, manufacturing, cities, and transportation.

Environmental Determinism

Geographers have repeatedly stressed environmental problems or shortages of certain raw materials as impediments to industrial development in the Soviet Union and in some other countries. Even George Cressey, who emphasized the dynamic transformation of the Soviet Union, stated that "permanent environmental restrictions . . . will never permit this part of the world to achieve strictly first-class rank."[7] "Never" is a long time! Who can predict the future course of technology, of discovery or utilization of new resources, of the rise of new industries? When I was a student in the 1930's many geographers and others were of the opinion that neither the Soviet Union nor Japan could become a major industrial power. Consider this quotation from a book on the economic geography of Europe by two professors of geography published in 1931: "The great mass of the Russian people are exceptionally poor . . . Yet, there has been little else than agriculture to which they could turn, as most of Russia is relatively remote from the sea and from foreign markets, and the lack of local varied resources has discouraged diversification."[8] This statement utterly misses the essentially internal nature of Soviet development, the richness of natural resources in the Soviet Union, and the importance of national policy and human resources in industrial development.

The best geographical analysis of the industrial potential of Japan in 1930 was John E. Orchard's monograph, Japan's Economic Position; the Progress of Industrialization. Orchard concluded that "her possibilities for industrialization are limited and there seems to be no prospect that Japan can attain a position of major importance as a manufacturing nation."[9]

---

[7] George B. Cressey, How Strong Is Russia? A Geographic Appraisal (Syracuse, N.Y.: Syracuse University Press, 1954), p. v.

[8] W. O. Blanchard and S. S. Visher, Economic Geography of Europe (New York: McGraw-Hill Book Co., 1931), p. 371.

[9] John E. Orchard, Japan's Economic Position; the Progress of Industrialization (New York: Whittlesey House, McGraw-Hill Book Co., 1930), p. 482.

But the Soviet Union and Japan now rank second and third among the industrial powers in the world, surpassing Britain and Germany in many types of production. Industrialization depends not only on richness of raw materials but also upon markets, manpower, and development of industrial capital, equipment, and skills. A large disciplined intelligent labor force is more important than the physical presence of a single resource, such as coal or iron. As Lenin observed, "The primary productive force of all humanity is the worker."[10]

Ellsworth Huntington had another type of environmental determinism, climatic.[11] Where cyclonic storms were frequent and variability of the weather was pronounced, man's brain was stimulated, and hence civilization rose to its highest levels in middle-latitude marine climates. Russia with its long dull winter was doomed!

The "Normal" Year

For a long time after World War I, British geographies presented data from before that war as being more nearly "normal." The same thing happened for a while after World War II. Geographers in order to study in detail distributional patterns are likely to select a typical or "normal" year and perhaps unconsciously assume that conditions fluctuate about that norm and thus neglect changes over time. Under conditions of relative stability, as in agriculture in a peasant society, where fluctuations or growth are slight, such methods are reasonably satisfactory. For industrial geography in a rapidly developing country, however, the best concept is not the typical or normal year but rather that of a point on a rising curve. Such is the case in the industrial output of the Soviet Union or of Japan in recent years. In the 25 years between an article I wrote on Soviet cities and a book on the same subject, that is between 1945 and 1970,[12] the urban population of the Soviet Union grew from about 65 million to over 136

---

[10] Vladimir Il'ich Lenin, Polnoe Sobranie Sochinenii, Vol. 38, p. 359, as quoted in V. V. Pokshishevskii, Geografiia Naseleniia SSSR: ekonomiko-geograficheskie ocherki (Moskva: Izdatel'stvo "Prosveshchenie," 1971), p. 6.

[11] Ellsworth Huntington, Civilization and Climate (New Haven: Yale University Press, 1915), 333 pp. and Ellsworth Huntington and Sumner W. Cushing, Principles of Human Geography (New York: John Wiley, 1921), pp. 248-259.

[12] Chauncy D. Harris, "The Cities of the Soviet Union," Geographical Review, Vol. 35, No. 1 (January, 1945), pp. 107-121, and Cities of the Soviet Union: Studies in Their Functions, Size, Density, and Growth (Washington, D.C.: Association of American Geographers, 1970), 484 pp. (Monograph Series No. 5).

million, or more than double, while the production of steel rose 9-fold from 12 million metric tons to 116 million tons, of petroleum production some 18-fold from 19 to 353 million metric tons, and of natural gas some 60-fold from 3 to 200 trillion cubic meters.[13] The rise in urban population and industrial production in Japan in this same period has been equally dramatic.

Soviet geographers have been somewhat more perceptive than their Western counterparts in recognizing the sense of the dynamic evolutionary character of the economic geography of industry. I recall the book by the Soviet geographer and writer, N. N. Mikhailov, Soviet Geography; the New Industrial and Economic Distributions of the U.S.S.R.,[14] with an introduction by Sir Halford Mackinder, published in 1935, which was roundly criticized at the time because it seemed to fail to distinguish among what existed, what was in course of development, and what was merely conceived or planned for some future period. The criticism was justified, yet the viewpoint that more attention needed to be paid to unfolding developments and plans was vindicated by the subsequent evolution of the economic geography of the Soviet Union.

"Everything Is in English "

It has been asserted by scholars of some repute that in some fields all important literature is available in English. Whether this assertion is true for some fields, and I seriously doubt that it is, it clearly is utterly fallacious in most area-studies fields. It could be asserted that all important literature is in French, for all bibliographies in the Grand Larousse Encyclopédique include only publications in the French language, suggesting, I suppose, that there is no significant literature on the United States or Great Britain in English, or on Germany in German, or on the Soviet Union in Russian. Nevertheless it must be clear that the bulk of the serious work on any major country will be in the language of that country. The principal current geographical bibliographies produced in the Soviet Union, Germany, France, Great Britain, and the United States, all agree on the overwhelming dominance of the language of a country in the cited literature on that country. Thus of the entries in the Bibliographie

---

[13] Gosudarstvennyi Piatiletnii Plan Razvitiia Narodnogo Khoziaistva SSSR na 1971-1975 gody (Moskva: Izdatel'stvo Politicheskoi Literatury, 1972), p. 7.

[14] N. N. Mikhailov, Soviet Geography: the New Industrial and Economic Distributions of the USSR (London: Methuen, 1935), 232 pp.

Géographique Internationale in 1969, 92 per cent of those on the Soviet Union are in Russian, 94 per cent of those on Germany are in German, 90 per cent of those on the British Isles are in English, and 96 per cent of those on France are in French. For all other countries of Europe, taken as a group, 62 per cent of the works listed are in languages other than English, French, German, or Russian, in a great variety of national languages, such as Italian, Spanish, Polish, or Dutch. These are all areas with long traditions of scholarship and a substantial body of national scholars. For some areas of the world, the bulk of the scholarly works have been written by outsiders. The Caribbean and the Pacific Islands are two cases in point. For the Middle East, Africa, Latin America, and India the situation is intermediate with both a substantial fraction of indigenous work and major contributions by outsiders.

Of the serious scholarly and scientific literature on geographic problems in the Soviet Union, I would estimate that more than 90 per cent is in Russian, Ukrainian, Estonian, Latvian, Lithuanian, Georgian, Armenian, or some other language of the USSR. There are about 10, 000 active geographers in the Soviet Union working on geographic problems largely in that country and not more than about 100 geographers from all other countries who make original studies of geographical topics in the Soviet Union. Fortunately there is a substantial corpus of material available in English, including a large volume of articles translated from Russian works by Soviet authors. The journal Soviet Geography: Review and Translation, for example, has published translations of articles by more than 500 different Soviet geographers or scholars in related disciplines.

I mention once again five pitfalls in the study of foreign areas: (1) the danger of attributing to another civilization characteristics of one's own culture without evidence from the country in question; (2) the deceptive insights of grand generalizations; (3) the non-historical and culturally insensitive blindness of environmental determinism; (4) the inadequacy of the "normal" years; and (5) the linguocentric myopia of assuming that everything is in English, or any other language for that matter.

I now turn to some critical needs for effective work on or understanding of the Soviet Union.

The Need for Statistics, Maps, and Field Work

At various periods in the study of the Soviet Union the geographer has been greatly impeded by the lack of statistics. From the mid-1930's to the mid-1950's, for example, there was a prolonged drought on statistics of all kinds of interest to geographers--of population or its distribution, and of agricultural or industrial production. Then in 1956 appeared the first statistical yearbook in a score of years. Since then the annual statistical yearbooks have improved enormously and this year even the concrete figures of the 7th five-year plan for 1971-1975 were published. However, the census materials are still far inferior to older ones: the 1939 census was never published, the 1959 census was scanty in data compared with the 1926 census or the census of 1897; but there is hope that the 1970 census volumes may yet provide more detail.

The absence of data causes two contradictory tendencies, both of which lower the scholarly level of work. On the one hand, there may arise a general quantitative vagueness and inexactness; on the other hand, scholars may spend an excessive proportion of their time in trying to arrive at rough figures by extrapolation, interpolation, projections, or dubious statistical manipulations and thus become embroiled in the process of trying to get figures to the neglect of the proper geographical analysis of the phenomena under study. One economist who has spent much of his life in trying from afar to construct data from spotty sources confessed that he simply would not know what to do if he were suddenly free to make a study inside the Soviet Union.

Among Soviet geographers a similar blind spot tends to develop in the absence of accurate detailed maps. Studies of urban patterns within Soviet cities by Soviet geographers have a certain vagueness and inexactness because, in the absence of detailed city maps, students tend to lapse into generalized descriptions that reveal a lack of precision in thinking about, analyzing, or describing the internal structure or patterns of cities.

A major weakness of geographic studies by foreigners of the Soviet Union is the lack of opportunity for detailed concrete field investigations with observation, mapping, recording, and interviewing. Oddly enough, Ellsworth Huntington, not generally thought of as a field observer, is one of the very few foreigners ever to have made detailed field studies in the area, but that was at the beginning of the century in Middle Asia. The ability to do field work has apparently atrophied from disuse among geographers specializing on the Soviet Union and if opportunities suddenly were to open up, it is not certain how effectively they would be utilized, at least for a while.

The problems of statistics, maps, and field work in geographic or other investigations are not limited to the Soviet Union. In the last two decades the People's Republic of China has posed each of these problems in an even more difficult form. The trend of the last decade has been a substantial improvement in the availability of statistics, often reported through international agencies such as the United Nations, for a diminution in the availability of maps because of military restrictions in areas such as India and Pakistan or parts of the Middle East, and persistent signs of a reduction in the freedom of foreigners to do unrestricted field work in many of the less developed countries now encouraging a strong nationalism in academic policy. Access to the field may be dependent on the scholar working on problems conceived by national leaders to be of direct value in the planning or development of a particular country, or on his involving personnel from the area in training in order that a local cadre of scholars and scientists and an infrastructure of local institutions can be built up more rapidly, or on his avoiding certain types of topics, such as minority groups, or opposition political movements, or industries deemed to have military significance. Or certain materials, as in archives, may not be made available until after they have been utilized by local scholars. It appears likely that in the decade ahead increasing restrictions will be placed upon the work of outside scholars in many countries.

The Need for Comparisons and Contrasts

One needs to see a country or political grouping or cultural area in the perspective of comparisons with other areas. Comparisons are like stereo pairs of aerial photographs; they show perspective, bring objects into relief, and highlight outstanding features. The opposite is the approach of the antiquarian who knows everything about a very small topic but has no sense of proportion, significance, or comparison.

For the general world setting of the Soviet Union, I have found comparison with North America helpful to students. Both are of similar size, about 22 million square kilometers (or 8 1/2 million square miles for those who have not yet moved intellectually into the international metric system). Both have a population of the order of 250 million persons. In each case, one language group, English and Russian respectively, constitutes more than half the population, but in each case other ethnic groups of high literacy and culture are also present, French Canadians and Spanish-speaking Mexicans in North America, Ukrainians, Estonians, Latvians, Lithuanians, Georgians, Armenians, and many other nationalities in the Soviet Union. In both there are also many small indigenous groups, speaking a variety of tongues, such as Indians and Eskimos in North America or peoples of the North in the Soviet Union.

Both have a core of fertile agricultural land bounded by vast areas too cool or too dry for ordinary agriculture, the area in the United States and Canada east of the 100th meridian and south of the Canadian shield, bounded on the north by the vast coniferous forests of Canada and Alaska and on the west by semiarid and arid lands, especially of the United States and Mexico; in the Soviet Union the fertile triangle with corners on the Baltic and Black seas and extending east to a southern strip in Western Siberia bounded on the north by the taiga, the greatest stretch of forest land on the face of the globe, and on the southeast by the semiarid dry steppes and the deserts of Soviet Middle Asia.

The Soviet Union is oriented east-west, while North America extends more north-south. Thus the Soviet Union has a more continental climate, indeed the most continental of any part of the globe. The Soviet Union also lies somewhat farther north; the southern border of the eastern half of the country lies at a latitude of 50 degrees, or about the Canadian-United States boundary, and only a small part of the western half of the country lies south of this latitude. Only a tiny segment of the country in the deserts of Middle Asia lies as far south as Chicago (42° north latitude). As a result, the agricultural lands of the Soviet Union are generally cooler and drier than in North America and thus agricultural productivity is generally substantially less, with lower yields per hectare, poorer possibilities for increasing yields, higher variability of yields from year to year, a more limited range of crops that can be grown economically, and a greater dominance of grains.[15] Partly as a result of difficult physical conditions but partly also because of a long period of very low agricultural prices and of certain organizational problems, agricultural production thus far in the Soviet Union has lagged behind planned production, behind other sectors of the Soviet economy, and also behind increases in agricultural production in Western Europe or the United States. The agricultural lands of the Soviet Union can be best compared with the spring wheat belt of the northern Great Plains of the United States and the prairie provinces of Canada, with some areas similar to the winter wheat belt. The Soviet Union lacks large agricultural belts with both high humidity and high temperatures characteristic of the American corn belt or cotton belt. Both countries have extensive irrigated oases amid deserts or semideserts with specialty crop production including cotton. It is useful to note that about half of the Soviet Union is underlain by permafrost, permanently frozen subsoil in which only a thin surface layer thaws in the short summer.

---

[15] Chauncy D. Harris, "Soviet Agricultural Resources Reappraised," Journal of Farm Economics, Vol. 38, No. 2 (May, 1956), pp. 268-273.

North America and the Soviet Union, both very large areas, possess rich mineral resources of fuel such as coal, petroleum, and natural gas, and of metalliferous ores of iron and copper.

There are some interesting contrasts also. Take the revolutions that established the government of the United States from the British colonies in North America and the government of the Soviet Union from the Russian Empire.[16] The American Revolution of 1776 was characterized by a mistrust of unchecked centralized government power. As a result, in the United States political responsibility and power are divided among the executive, legislative, and judicial branches of government at the federal level, and among federal, state, and local governmental units. In addition, many aspects of society are held altogether outside the domain of government as expressed in the separation of church and state and the guarantee of civil liberties in the bill of rights.

The Communist Revolution of 1917, in sharp contrast to the American Revolution of 1776, was developed on the basis of great faith in the power of the government to transform society for the betterment of man. The Communists seized the government not to restrict its power but rather to use it. They aspired fundamentally to remold society through a state monopoly of the production of goods, of means of communications, of education, and of science.

The great modernizing revolutions that took place much earlier in the West and at a more leisurely pace of gradual transition came concurrently in the Soviet Union and thus took a more dramatic and radical form: the religious revolution, which in the West found expression in the Reformation, the Counter-Reformation, and the gradual secularization of most aspects of life, took the form in the Soviet Union of militant atheism; the economic revolution, which as the Industrial Revolution extended over more than a century in the West took the form in the Soviet Union of state ownership and management of the entire economy to promote social goals and to speed industrialization; the democratic political revolution, which found expression in the West in the American and French revolutions and the gradual diffusion of political power in Britain through a series of reforms, in the Soviet Union took the form of transfer of power from the autocracy of the Tsars to the dictatorship of the proletariat through the leadership of the Communist party with centralized authority but democratic participation; the intellectual revolution, which in the West flowered in the Age

---

[16] Chauncy D. Harris, "Society, Science, and Education in the Soviet Union," *Proceedings of the National Academy of Sciences*, Vol. 45, No. 5 (May, 1959), pp. 684-692.

of Reason or the Die Aufklärung, in which faith developed in the perfectibility of man and of social institutions, in the ability of man by rational thought and scientific investigation to improve himself and society and to rule the universe, in the Communist Revolution took the form of an optimistic faith in the ability of the Party and the Government through science and industrialization to transform society and social relations and to create a rational communistic world order. Thus these four great modernizing revolutionary changes, the Reformation, the industrial revolution, the democratic revolution, and the Enlightenment, have touched both the West and the Soviet Union but at different times and at different rates and thus have been expressed in different forms.

Another comparison might be made between the political philosophy guiding the Communist parties of the Soviet Union and the People's Republic of China. On two very important lines they differ radically. First on the professionalization of the party apparatus. One of Lenin's great contributions to the success of the communist revolution in the Soviet Union was his diagnosis of the failure of the 1905 uprising as resulting in part from the lack of a disciplined leadership and his subsequent development of full-time professional disciplined party leaders, who in the critical days of 1917 enabled a tiny minority to seize power. The party bureaucracy to this day plays a leading role in the determination of party policy and governmental programs. In contrast, the single most striking attribute of the thought of Mao Tse-tung that impressed me from a reading of his little red book is his profound distrust of any bureaucracy, whether governmental or party, and his successive destructions of the party leadership structure. Of course, Stalin in his day also wiped out great chunks of the leadership of the party. But at least philosophically Lenin put the responsibility for political leadership directly on the party organization, specifically on the central committee of the party, and Mao Tse-tung placed emphasis on a continuing revolutionary fervor of the masses that would control party leaders. One other sharp contrast is between the urban-oriented Soviet party which would like to make the countryside like the city with the development of an agricultural city, or agrogorod, and the transformation of the peasantry into an industrial proletariat, whereas the Chinese leaders would rather make the city like the country and take industries into the countryside. The course of communism in the Soviet Union was in part the conquest of the "lumpen mass" of the peasantry by the urban proletariat, but in China the revolution was the siege of the city by the agrarian-based communist movement. These two contrasts in the essence of the communist movements in the two greatest communist powers are only elements in their differences, which also have deep roots in nationalism and rivalry.

The Soviet Union also presents an interesting contrast with Latin America or the Middle East. The USSR has political and economic unity but cultural, religious, ethnic, and linguistic diversity, whereas Latin America is fragmented politically and economically but has a considerable degree of linguistic and cultural unity through the use of Spanish and religious unity in the Roman Catholic Church, though under the veneer of Spanish culture there are major Indian elements, particularly in Mexico, Guatemala, Peru, and Bolivia. In both the Soviet Union and in Latin America the major languages, Russian and Spanish, are expanding rapidly, particularly against the smaller indigenous languages, just as English and French are becoming more widespread as languages of wide communication in Africa. In the Middle East as in Latin America there is political fragmentation but over wide areas also linguistic affinity through the use of Arabic and religious affinity through Islam.

India and the Soviet Union are the two major countries with the greatest ethnic and linguistic diversity. In each the basic political units in a federal system are principally ethnic and linguistic, and the unity of the country poses difficult political problems, particularly in India. The position of Russian, however, in the Soviet Union, though it is the mother language of only half the inhabitants, is much stronger than of any language in India, where English is spoken by only a very thin veneer of the educated. Hindi is spoken by only about a third of the population and has never been the principal language of government, education, business, journalism, or the courts; after all it was Persian that English replaced as the governing language in north India. By way of contrast Russian has always been the language of the dominant group in the Russian Empire and the language of government, commerce, education, and science. Historically as well as currently, as Mayakovskiy observed, the Russian state emanates from the Kremlin.

The Role of Scholars from within the Area and of Those from Outside

What then can scholars from outside the area contribute? For any major world area, or civilization, or country, obviously most of the study will be made by the geographers from the area itself.

It is often difficult for those within an area to generalize about it as a whole. They are more engaged in detailed and specialized research on highly specific topics or on small regions. Thus in spite of the high level of geographic research in Germany and in the United States, a good general geography of either area treated as a whole by geographers from within is lacking, though there are regional treatments what describe the regional subdivisions in some detail.

On the other hand, the scholars from outside an area often will not have the opportunity, financial support, time, preparation, or access to make detailed field studies of interesting problems within an area. Furthermore many problems of an applied or practical nature can best and in many cases only be done by residents of the area, who have specific responsibility within state or research organizations to make studies directly relevant to the solution of immediate practical problems or for specific projects in various stages of planning, construction, or utilization.

There sometimes arises a complete separation of level of discourse between detailed work within a country, for example the study of the utilization of a particular dam, and global generalizations from abroad, without any relationship between the two.

What I wish to suggest is that the most fruitful work should come from an interaction between detailed scientific investigations and generalizations at the level of an entire country, that the latter should be solidly based on the former, taking account, of course, of appropriate scale of generalization. The general treatises on the larger topics will be undertaken by both native and foreign geographers, those by the former typically more solidly based on the corpus of original investigations within the country, the latter having possibly more insight from comparisons with other areas. But the former need the depth which comes from knowledge of other areas, for who would know his own country must know others also, and the latter need to be based on a solid and profound knowledge of the detailed research within and in the language of the country. Thus there is need for an intimate interaction, a dynamic dialectic, a creative collaboration.[17]

---

[17] For an account of work by American geographers on the Soviet Union see W. A. Douglas Jackson, "Geography," in American Research on Russia, ed. by Harold H. Fisher (Bloomington, Indiana: Indiana University Press, 1959), pp. 113-122.

For accounts of Soviet work in geography see:

Soviet Geography: Accomplishments and Tasks. A Symposium of 50 chapters, contributed by 56 leading Soviet geographers and edited by a committee of the Geographic Society of the U.S.S.R., Academy of Sciences of the U.S.S.R., I. P. Gerasimov, chairman. English edition edited by Chauncy D. Harris (New York: American Geographical Society, 1962), 409 pp. (Occasional Publication No. 1).

Soviet Geography: Review and Translation (American Geographical Soci-

In summary, I have tried to emphasize three needs for effective work on or understanding of the Soviet Union: (1) The requirements for statistics, maps, and field work; (2) The insights that may be suggested by comparisons and contrasts with other areas, realizing, of course, that such insights suffer from the potential dangers of the grand generalizations; and (3) the somewhat differentiated roles of scholars from within and of those from outside the area and the importance of their interaction.

---

ety), New York. 1960-. Monthly except July and August. Theodore Shabad, Editor and Translator.

Abraham Melezin, "Trends and Issues in the Soviet Geography of Population," Annals of the Association of American Geographers, Vol. 53, No. 2 (June, 1963), pp. 144-160.

Roland J. Fuchs, "Soviet Urban Geography: An Appraisal of Postwar Research," Annals of the Association of American Geographers, Vol. 54, No. 2 (June, 1964), pp. 276-289.

George J. Demko, "Trends in Soviet Geography," Survey; A Journal of Soviet and East European Studies, No. 55 (April, 1965), pp. 162-170.

Dora Fischer, "Siedlungsgeographie in der Sowjetunion," Erdkunde; archiv für wissenschaftliche Geographie, Vol. 20, No. 3 (September, 1966), pp. 211-227.

R. A. French, "Historical Geography in the USSR," Soviet Geography: Review and Translation, Vol. 9, No. 7 (September, 1968), pp. 551-561.

Robert A. Lewis, "The Study of Postwar Internal Migration in the USSR," Soviet Geography: Review and Translation, Vol. 10, No. 4 (April, 1969), pp. 157-166.

Raymond S. Mathieson, "The Soviet Contribution to Regional Science: A Review Article," Journal of Regional Science, Vol. 9, No. 1 (April, 1969), pp. 125-140.

Robert G. Jensen and Gerald J. Karaska, "The Mathematical Thrust in Soviet Economic Geography; Its Nature and Significance," Journal of Regional Science, Vol. 9, No. 1 (April, 1969), pp. 141-152.

Robert G. Jensen and Gerald J. Karaska, "Application of Mathematical Methods in Soviet Economic Geography," Soviet Geography: Review and Translation, Vol. 10, No. 9 (November, 1969), pp. 501-506.

Paul E. Lydolph, "Soviet Work and Writings in Climatology," Soviet Geography: Review and Translation, Vol. 12, No. 10 (December, 1971), pp. 637-666.

Harold D. Foster, "The Changing Focus of Geomorphology," Soviet Geography: Review and Translation, Vol. 13, No. 6 (June, 1972), pp. 337-343.

R. A. French, "Historical Geography in the USSR," Progress in Historical Geography, ed. by Alan R. H. Baker (London: David and Charles; New York: Wiley-Interscience, 1972), pp. 111-128.

For an earlier period see David J. M. Hooson, "The Development of Geography in Pre-Soviet Russia," Annals of the Association of American Geographers, Vol. 58, No. 2 (June, 1968), pp. 250-272.

For comparisons of Soviet and American geography see David Hooson, "Phases in the 20th Century Development of Russian and American Geography: A Comparison," Proceedings of the Association of American Geographers, Vol. 1 (1969), pp. 66-69.

CHAPTER 2

LATIN AMERICA

James J. Parsons
University of California, Berkeley

Among the great world culture areas Latin America stands unique, in many ways still a New World, largely unknown and unstudied. The extraordinary fact of its discovery, and the infinitely complex process of its absorption into the larger world, begun on the shores of Hispaniola in 1492 and still in process, represents one of the great epics of human experience. The origin and character of the indigenous societies of America, previously unknown and undreamt of, as well as the proper relationship of European society to them, have provided the basis for intense philosophical speculation and scholarly investigation for generations of Western scholars.[1] Latin America stands alone, too, as the one major world area in which effective racial and cultural mixing has been achieved. Its 200 million peoples represent in large part a quite recognizable new race and culture, however diverse their political and social components. But Latin America is also the home of major Indian survivals, even integrated cultural entities, and these stubbornly conservative groups themselves have importantly influenced the nature of the dominant society. Here exists a unique blend of the European and native cultures, the indigenousness of which remains largely unstudied. The Catholic church, the Spanish and Portuguese languages, and Iberian institutions, even the flattening steamroller called modernization, in many areas but thinly veneer the persistent and deep roots in the aboriginal and colonial American past.[2]

Interest in these contrasting cultures and economies of Latin America, in

---

[1] The most recent review is J. L. Abellán, *La idea de América: origen y evolución* (Madrid, 1964).

[2] This theme has been provocatively developed by S. J. and B. H. Stein, *The Colonial Heritage of Latin America: Essays on Economic Dependence in Perspective* (Oxford: Oxford University Press, 1970). See also, R. F. Crist, "The Latin American Way of Life," *American Journal of Economics and Sociology*, Vol. 27 (1968), pp. 63-76.

complex spatial and temporal juxtaposition, is further heightened for geographers by the extreme diversity of the physical environment they inhabit, from the rainless Atacama to the Amazonian rain forest and to the highest inhabited mountains on earth--climatically only the extreme continentality of the northern hemisphere mid-latitudes is missing. This richness of cultural and environmental dimensions is heightened by the availability of an extraordinary documentation in European languages, much of which has been published and the rest of which is preserved in well-organized and preserved archives. Further, and increasingly, Latin America is in a sense a part of us--in the Hispanic-American Southwest, in California, in the Cuban colonies of Florida and the Puerto Rican barrios of Manhattan--with clear relevance for the understanding of the contemporary society of the United States. Finally, for North Americans, these foreign lands directly to the south are uniquely accessible, and often at minimal cost. No cultural boundary on earth offers more striking contrasts than those on either side of the United States-Mexican frontier between the mouths of the Rio Grande and the Tijuana rivers, and few are easier to cross. Only the peculiar Anglo-Saxon block against foreign languages, coupled at times with a curious and possibly growing inhibition among geographers against trying to know things first hand, of getting away from the difficulties of time and place, has kept more of our kind from pursuing their studies in these extraordinary lands.

If one were designing on a _tabla rasa_ an ideal but imaginary New World it would be difficult to conceive, even in one's wildest dreams, of a set of human and environmental circumstances more extraordinary, more original than that offered by the Americas. The English and French speaking lands north of the Rio Grande, much later occupied, were to become firmly attached to the Old World as economic and cultural extensions of Europe. To the south, in Ibero-America, a very different society or societies emerged--at the same time monolithic and plural--strongly influenced by aboriginal cultures, by the dominant tropicality of the climate, by the existence of extraordinary metallic riches, and by the historical accident of the political dominance of two strongly Catholic and centralist European powers.

Latin America is almost an island in the world ocean, standing in contrast and isolation from the other great world island--the Old World--which houses the other areas under consideration in this symposium (the Caribbean excepted) in one contiguous block. Distance and isolation have played a very real role, often not fully appreciated, in its evolution. Even today it stands at one side from the great avenues of world commerce. The links and strands connecting these two worlds, the interchange of plants and animals, ideas, and institutions provide us with one of the grand and continuing themes of scholarly investigation.[3] Such isolation, combined with unique cultural and ecologic circumstances, offers unparalleled opportunities for comparative studies, as yet but little exploited.

North American Geographers in Latin America

The history of the involvement of North American geographers in Latin American studies dates back to William Morris Davis, Mark Jefferson, George M. McBride, and Isaiah Bowman.[4] Bowman, as head of the American Geographical Society, gave that organization a special role in Latin American research, culminating with the Millionth Map of Hispanic America project, completed in 1944. Following World War I, Carl O. Sauer and Preston James, along with Clarence Jones and Robert Platt, vigorously entered into the Latin American field, Sauer in historical geography, the others primarily on the contemporary regional and economic side. Sauer's extensive bibliography, with more than 15 papers on Latin America, is to be found in John Leighly (ed.), Land and Life: Selected Writing of Carl Ortwin Sauer (Berkeley and Los Angeles, 1963); James' bibliography appears in Donald Meinig (ed.), On Geography: Selected Writings of Preston E. James (Syracuse, 1971). It is noteworthy that both of these men continue today as active and highly original contributors to our literature. Their impact in shaping the field has been great.

The following pages, based in part on earlier reviews,[5] are concerned largely with what North American geographers have done, are doing, and might do in Latin America area studies. It largely omits reference to the important work being done by Latin Americans themselves, especially in Brazil, Argentina,

---

[3]See, for example, C. L. Riley, J. C. Kelley, C. W. Pennington, and R. L. Rands (eds.), Men Across the Sea: Problems of Pre-Columbian Contacts (Austin: University of Texas Press, 1970). This important symposium contains contributions by several geographers.

[4]Preston James, "Studies of Latin America by Geographers in the United States," pp. 1-12 in Barry Lentnek, Robert Carmin, and Tom Martinson (eds.), Geographic Research on Latin America: Benchmark 1970 (Muncie, Ind.: Ball State University, 1971), 434 pp.

[5]James J. Parsons, "The Contribution of Geography to Latin American Studies," pp. 33-85 in Charles Wagley (ed.), Social Science Research in Latin America (New York: Columbia University Press, 1964); "Ecological Problems and Approaches in Latin American Geography," pp. 13-32 in Barry Lentnek et al. (eds.), Geographic Research on Latin America: Benchmark 1970 (Muncie, Ind.: Ball State University, 1971).

Chile, and Mexico, as well as the extensive French and German geographical literature. The latter, especially, has been in good part a by-product of joint programs between Latin American and European governments or universities, as exemplified by the Mexican-German scientific project currently under way in the Puebla-Tlaxcala area. In contrast, and with a few conspicuous exceptions, field research by North American visitors to Latin America has been by individuals working alone and independently.

However, North American geographers working on Latin America have recently been coming together. The proceedings of the first Conference of Latin Americanist Geographers (CLAG), held at Muncie, Indiana, in May 1970, contain some 40 review papers, focused on different aspects of geography and accompanied by often extended bibliographies (Lentnek, Carmin, and Martinson, eds., 1972). A second conference, held at Syracuse in 1971 (Proceedings in press), focused on geography's role in regional development in Latin America. A third is scheduled for Calgary in 1973. An earlier review, "The Contribution of Geography to Latin American Studies," prepared by the author, appeared in the volume on Social Science Research in Latin America edited by Charles Wagley (Columbia University Press, 1964), the product of an interdisciplinary conference at Palo Alto. The background of this paper underscores geography's "poor relation" position vis-a-vis the other social sciences in the Latin American field. Although the conference was organized by subject matter, no geographer was asked to contribute a paper. I was, however, an invited participant. No discussion of geography had been envisaged simply because "no one had thought of it." An invitation was quickly extended to me to prepare a position paper comparable to those for the other fields--but only after the termination of the meetings.

Pre-Historical and Historical Geography

Work in Latin America by English-speaking geographers has been strongly dominated by the historical approach. The lines of history are here deeply etched on the land and people, and an understanding of the present is heavily dependent on the past.[6] More than one geographer, with an original intent to do a conventional--or unconventional--contemporary field study, has been lured

---

[6]David J. Robinson, "Historical Geography in Latin America," pp. 168-186, 260-273 in A. R. H. Baker (ed.), Progress in Historical Geography (New York: Wiley, 1972). This fine survey contains an extraordinary wealth of bibliographic citations, especially from sources within Latin America.

irrevocably to the pre-Columbian or colonial past once he has confronted the Latin American reality, with its strong and diverse cultural survivals, and the impressive archeologic remains and archival documentation that makes historical study and elucidation both feasible and compelling. The extent of the historical bias is documented in a recent bibliography of Latin American historical geography.[7] It shows, too, the very strong predominance of work on Spanish-speaking areas as compared to that on Portuguese-speaking Brazil. This in part reflects the extraordinary attraction to scholars of the problems associated with the origins of Meso-American and Andean high cultures and the persistence of aboriginal influences--language, customs, ways of life--particularly in Mexico, Guatemala, Ecuador, Peru, and Bolivia. This bias is also explained in part by the low position and prestige of Portuguese language study in our schools, and, in the case of Mexico and Central America, the obvious ease of access to the North American scholar. Finally Brazilians themselves, much more than Spanish Americans, have been oriented to the present, to problems of modernization and development, despite the lateness of their independence from European control. The bibliography just mentioned, which seems representative of English language research, contains some 446 items. Of these 214 are localizable in Mexico and Central America, and 165 in South America, of which only 20 are in Brazil proper.

Geographical work on Latin America often tends to fuse with that of cultural anthropologists and historians as in community studies and in studies of the origins, spreads, limits, and modifications of culture complexes or traits. The links with anthropology and with anthropological method are in part a result of that subject's greater dependence on field work and direct observation than characterizes the other social sciences. The data of physical geography, soils, land forms, climate, and climatic change, are increasinly relied upon to resolve the knotty problems of past habitats and societies as the so-called ecological approach has given new vigor and direction to that profession and its handmaiden, archeology. The door is increasingly open to geographers for collaboration. The first volume of the <u>Handbook of Middle American Indians</u>, edited by Robert West, and the contributions in volume 6 of the <u>Handbook of South American Indians</u> on physical geography and biogeography are examples of the recognition by anthropologists of the role of the physical environment in human affairs,

---

[7]William M. Denevan, <u>A Bibliography of Latin American Historical Geography</u> (Washington, D.C.: U.S. National Section, Pan American Institute of Geography and History, Special Publication No. 6, 1971).

something many geographers themselves paradoxically seem willing to ignore or even deny.

A significant contributing factor to this emphasis on cultural and historical geography has been the strong and persistent influence of the maestro of Latin American historical geography, Carl Sauer of Berkeley, whose work with that of his students comprises a very significant part of this literature, virtually all of it on Spanish-language areas. The Ibero-Americana monograph series at the University of California, founded by Sauer and the anthropologist Alfred Kroeber in 1932, has been the vehicle for much of this work. Of its 51 volumes to date, some 16 have been by professional geographers, the rest by colleagues working on the fringes of the field with an active concern for human ecology and demographic or cultural history. A comparable number of monographs in the University of California Publications in Geography have been on Latin American or Caribbean topics, while several other Berkeley studies have had more limited distribution as Office of Naval Research reports.

Considering the remarkable consistency with which Sauer's ideas or "hunches" have been verified in time by archeology and plant genetics, it is perhaps curious that no more geographers have been attracted into the lines of prehistoric inquiry that he pioneered. His wide-ranging contributions have included studies on the antiquity of man in the Americas, pre-Columbian population densities, the origins and dispersal of domesticated plants and animals, pre-Columbian contacts between the Old World and the New World, native agricultural systems, and the nature and extent of environmental deterioration through time.

It is a striking and singular fact that it has been the archeologists and anthropologists who have most seriously engaged themselves in developing hypotheses and evidence on man-land or culture-environment relationships in Latin America. They have, in fact, often tended to see in cultural ecology a new kind of magic, perhaps in partial reaction against a too rigid cultural determinism.[8]

---

[8] Betty Meggers, "Environmental Limitation on the Development of Culture," American Anthropologist, Vol. 56 (1954), pp. 801-824; Amazonia: Man and Culture in a Counterfeit Paradise (Chicago, 1971); Angel Palerm and Eric Wolf, Ecological Potential and Cultural Development in Mesoamerica (Washington, D.C.: Pan American Union Social Science Monographs III, Studies in Human Ecology, 1957); Robert L. Carneiro, "Slash-and-Burn Cultivation among the Kuikuru and Its Implications for Cultural Development in the Amazon Basin," pp. 47-67 in J. Wilbert (ed.), The Evolution of Horticultural Systems in Native South America (Caracas, 1961); Michael Coe and K. Flannery, "Microenvironments and Mesoamerican Pre-History," Science, Vol. 143 (1964), pp. 650-654; D. S. Byers (ed.), The Prehistory of the Tehuacán Valley, Vol. 1 (Austin: University of Texas Press, 1967); William T. Sanders and Barbara J. Price, Meso-America: The Evolution of a Culture (New York, 1968).

Past climates and climatic change, in particular, become increasingly relevant as we push back in time with our evidence, but work here by geographers[9] has not yet resulted in any overall synthesis. What is needed, clearly, is a Karl Butzer for Latin America.

The anthropological literature on pre-historic land and livelihood in Latin America is probably more extensive than the totality of all geographic writing on the area. It has been conspicuously reported in journals like Science, Scientific American, Natural History, and in the popular press, and the work has been generously supported by funding agencies. It is not a fad and it is not a minor theme of only peripheral significance to the understanding of man's place in nature. It is in the mainstream of contemporary scientific inquiry. It involves the laws governing climate and landscape evolution, and it involves perception of human habitats, production systems, and settlement patterns and a host of other concerns close to the heart of any reasonably defined geography. Yet here, in our own back-yard, we continue to be outflanked and out-numbered.

Pre-historic and primitive agricultural production systems and techniques --milpa farming, irrigation, terracing, ridged fields--have perhaps most attracted our attention within this context.[10] Again, Sauer's early classic work

---

[9] Herbert Wilhelmy, "Die Klimamorphologie und Pflanzengeographie Entwicklung des Trockengebietes am Nordrand Südamerikas seit dem Pleistozan," Die Erde, Vol. 3 (1954), pp. 244-273; "Eiszeit und Eiszeitklima in den feucht tropischen Anden," Petermann's Mitteilungen, Ergänzungsheft 262 (1957), pp. 281-310; James Lahey, On the Origin of the Dry Climate in Northern South America and the Southern Caribbean (Cambridge, Mass.: Geophysics Research Directorate, Air Force Research Center, Scientific Report No. 10, 1958); C. J. Heusser, "Late Pleistocene Environment of the Laguna de San Rafael Area, Chile," Geographical Review, Vol. 50 (1960), pp. 555-577; O. Dollfus, "Cambios climáticos cuaternarios en los Andes peruanos," Boletín Soc. Geogr. Lima, Vol. 83 (1964), pp. 65-74; Vol. 84 (1965), pp. 75-83; W. A. Wood, "Recent Glacier Fluctuations in the Sierra Nevada de Santa Marta," Geographical Review, Vol. 60 (1970), pp. 374-392; and especially the numerous papers by Stephan Hastenrath, as e.g., "Certain Aspects of the Three-Dimensional Distribution of Climate and Vegetation in the Mountains of Central America and Southern Mexico," Colloquium Geographicum (Bonn), Vol. 9 (1968), pp. 122-130.

[10] E.g., Robert C. West, "Las chinampas de Mexico: poesia y realidad en los 'jardines flotantes,'" Cuadernos Americanos, Vol. 50 (1968), pp. 165-182; "Population Densities and Agricultural Practices in Pre-Columbian Mexico, with Emphasis on Semi-Terracing," Verhandlungen des XXXVIII Int. Amerikanisten Kongresses (Stuttgart-München, 1968), Vol. 2, pp. 361-369; L. C. Herold, Trincheras and Physical Environment along the Río Gavilán, Chihuahua (Dept. of Geography, University of Denver, Tech. Paper 65-1, 1965); James J. Parsons and William Bowen, "Ancient Ridged Field Agricultures in the San Jorge River Flood Plain, Colombia," Geographical Review, Vol. 56 (1966), pp.

in northwest Mexico[11] set the example for numerous later studies of pre-Columbian population densities and the thorny issue of carrying capacity and its relationship to cultural levels and dietary patterns.[12] The convergence of human geography with classical historical ethnography is exemplified by the work of several of his students.[13] Others[14] have made regional studies among modern

---

317-343; Gene C. Wilken, "Drained-Field Agriculture: an Intensive Farming System in Tlaxcala," Geographical Review, Vol. 59 (1969), pp. 215-241; "Food-Producing Systems Available to the Ancient Maya," American Antiquity, Vol. 36 (1971), pp. 432-448; William M. Denevan, "Ridged Field Agriculture in the Americas," Science, Vol. 169 (1970), pp. 647-654; Bernard Neitschmann, "The Substance of Subsistence," pp. 167-181 in Barry Lentnek et al. (eds.), Geographic Research on Latin America: Benchmark 1970 (Muncie, Ind.: Ball State University, 1971); David R. Harris, "Ecology of Swidden Cultivation in the Upper Orinoco Rain Forest, Venezuela," Geographical Review, Vol. 61 (1971), pp. 475-495; Ray F. Watters, Shifting Cultivation in Latin America (Rome: FAO Forestry Development Paper 17, 1971).

[11] Carl O. Sauer, "The Aboriginal Population of Northwest Mexico," Ibero-Americana, No. 10 (Berkeley, 1935).

[12] E.g., B. L. Gordon, "Human Geography and Ecology in the Sinú Country of Colombia," Ibero-Americana, No. 39 (Berkeley and Los Angeles, 1957); H. Homer Aschmann, "The Central Desert of Baja California: Demography and Ecology," Ibero-Americana, No. 42 (Berkeley and Los Angeles, 1959); William M. Denevan, "An Aboriginal Cultural Geography of the Llanos de Mojos of Bolivia," Ibero-Americana, No. 48 (Berkeley and Los Angeles, 1966); "The Aboriginal Population of Western Amazonia in Relation to Habitat and Subsistence," Revista Geográfica (Rio de Janeiro), Vol. 72 (1970), pp. 61-86; D. W. Lathrap, "Aboriginal Occupation and Changes in River Channel on the Central Ucayali, Peru," American Antiquity, Vol. 33 (1968), pp. 62-79; W. Borah and S. F. Cook, "Conquest and Population: A Demographic Approach to Mexican History," Proceedings of the American Philosophical Society, Vol. 113 (1969), pp. 177-183.

[13] E.g., Peveril Meigs, "The Kiliwa Indians of Lower California," Ibero-Americana, No. 15 (Berkeley, 1939); H. Homer Aschmann, op. cit. (1959); "Indian Pastoralists of the Guajira Peninsula," Annals of the Association of American Geographers, Vol. 50 (1960), pp. 408-418; Charles F. Bennett, "The Bayano Cuna Indians, Panama: An Ecological Study of Livelihood and Diet," Annals of the Association of American Geographers, Vol. 52 (1962), pp. 32-50; Campbell W. Pennington, The Tarahumar of Mexico: Their Environment and Material Culture (Salt Lake City: University of Utah Press, 1963); The Tepehuán of Chihuahua: Their Material Culture (Salt Lake City: University of Utah Press, 1969).

[14] Oscar Schmieder, "Settlements of the Tzapotec and Mije Indians, Oaxaca, Mexico," University of California Publications in Geography, Vol. 4 (1930); F. W. McBryde, Cultural and Historical Geography of Southwest Guatemala (Washington, D.C.: Smithsonian Institution, Institute of Social Anthropology, Publ. 4, 1947); Robert C. West, Cultural Geography of the Modern Tarascan Area (Washington, D.C.: Smithsonian Institution, Institute of Social Anthropology, Publ. 7,

Indian populations which may be conceptually indistinguishable from those of ecologically oriented anthropologists. Aschmann[15] has considered some of the problems associated with such work, not the least of which is the physical durability of the investigator. Among themes heavily dependent on Indian cultures that of plant and animal domestication is perhaps outstanding. Sauer's earlier work[16] has been continued by others.[17] The studies of Edwards[18] on aboriginal sailing craft on South America's west coast are of the same genre, as are land tenure investigations involving Indian systems.[19] Aschmann's perceptive analysis of the human ecology and demography of Baja California[20] provides an example of the importance of understanding not only the activities of individuals and communities but also the motivations and goals that induce them.

Although I have chosen to emphasize pre-Columbian and Indian studies, the colonial period in Latin America has offered themes of quite comparable interest for historical geographers. Pederson[21] on the Norte Chico of Chile and

---

1948); Donald Brand, Quiroga: A Mexican Municipio (Washington, D.C.: Smithsonian Institution, Institute of Social Anthropology, Publ. 11, 1951); William M. Denevan, "Campa subsistence in the Gran Pajonal, Eastern Peru," Geographical Review, Vol. 61 (1971), pp. 496-518.

[15]H. Homer Aschmann, "Indian Societies and Communities in Latin America," in Barry Lentnek et al. (eds.), Geographic Research in Latin America: Benchmark 1970 (Muncie, Ind.: Ball State University, 1971), pp. 129-137.

[16]Carl O. Sauer, Agricultural Origins and Dispersals (New York: American Geographical Society, 1952).

[17]Jonathan Sauer, "The Grain Amaranths: A Survey of Their History and Classification," Annals of the Missouri Botanical Garden, Vol. 37 (1950), pp. 561-632; Carl Johannessen, "The Domestication Process in Trees Reproduced by Seed: The Pejibaye," Geographical Review, Vol. 56 (1966), pp. 363-376; "The Domestication of Maize: Process or Event?" Geographical Review, Vol. 60 (1970), pp. 383-413; "Dispersal of Musa in Central America: The Domestication Process in Action," Annals of the Association of American Geographers, Vol. 60 (1970), pp. 685-699; David R. Harris, "The Origins of Agriculture in the Tropics," American Scientist, Vol. 60 (1972), pp. 180-193.

[18]E.g., Clinton Edwards, "Aboriginal Watercraft on the Pacific Coast of South America," Ibero-Americana, No. 42 (Berkeley and Los Angeles, 1965).

[19]Jean Borde and Mario Góngora, Evolución de la propriedad rural en el Valle de Puangue, 2 vols. (Santiago: Instituto de Sociología, Universidad de Chile, 1956).

[20]Aschmann, op. cit. (1959).

[21]Leland Pederson, The Mining Industry of the Norte Chico, Chile (Evanston: Northwestern University Studies in Geography, No. 11, 1966).

West[22] on the mining economy of northern Mexico and of northwestern Colombia are outstanding both as to method and content. Aschmann[23] offers an ingenious and original model for the evolution of a mine, with Chilean examples. The historical geography of agricultural crops and systems has had substantial attention,[24] as has the plantation economy, especially sugar,[25] but the definitive history of the plantation system--its origin in the Middle East and its transfer through the Canary Islands and Madeira to the New World remains to be written. Trade relations and the evolution of urban systems, too, have been subjects of increasing inquiry.[26] There are other historical geographies of political or administrative units,[27] of institutions such as the encomienda and the church, of

---

[22]Robert C. West, "The Mining Community in Northern New Spain: The Parral Mining District," Ibero-Americana, No. 30 (Berkeley and Los Angeles, 1949); Colonial Placer Mining in Colombia (Baton Rouge: Louisiana State University, Social Science Series, No. 2, 1952).

[23]H. Homer Aschmann, "The Natural History of a Mine," Economic Geography, Vol. 46 (1970), pp. 172-189.

[24]E.g., Henry J. Bruman, "The Culture History of Mexican Vanilla," Hispanic-American Historical Review, Vol. 28 (1949), pp. 360-376; A. S. Morris, "The Development of the Irrigated Agricultural Economy of Mendoza," Annals of the Association of American Geographers, Vol. 59 (1969), pp. 97-115; John Bergmann, "Distribution of Pre-Columbian Cacao Cultivation in Central America," Annals of the Association of American Geographers, Vol. 59 (1969), pp. 85-96; and the encyclopedic works of the Colombian scholar, Victor Manuel Patiño, Plantas cultivadas y animales domesticos en América equinoccial, 5 Vols. (Cali, 1963-70).

[25]J. H. Galloway, "The Sugar Industry of Pernambuco during the 19th Century," Annals of the Association of American Geographers, Vol. 59 (1969), pp. 285-304; Ward Barrett, The Sugar Hacienda of the Marqueses del Valle (Minneapolis: University of Minnesota Press, 1970).

[26]David L. Robinson, "Evolución en el comercio del Orinoco a mediados del siglo XIX," Revista Geográfica (Rio de Janeiro), Vol. 72 (1970), pp. 13-44; "Trade and Trading Links in Western Argentina during the Viceroyalty," Geographical Journal, Vol. 136 (1970), pp. 24-41. The more recent literature is reviewed in Howard Gauthier, "Geography, Transportation, and Regional Development," in Barry Lentnek et al. (eds.), Geographic Research on Latin America: Benchmark 1970 (Muncie, Ind.: Ball State University, 1971), pp. 333-342.

[27]E.g., Dan Stanislawski, "Tarascan Political Geography," American Anthropologist, Vol. 49 (1947), pp. 46-55; "The Political Rivalry of Pátzcuaro and Morelia; An Item in the 16th Century Geography of Mexico," Annals of the Association of American Geographers, Vol. 27 (1947), pp. 135-144; Peter Gerhard, A Guide to the Historical Geography of New Spain (Cambridge: Cambridge University Press, 1972).

native language survivals, of blacks[28] and other ethnic groups that have at best only been examined in preliminary form. Some of the more handsome and well preserved colonial towns--provincial centers such as Oaxaca, Camayagua, Popayan, Cajamarca, Sucre, and Ouro Preto--could provide attractive research themes that might throw illumination on modern problems of social persistence and change as well as providing pleasurable experience for the investigator.

The Physical Environment

From the time of Humboldt to the present the problems associated with the land forms, climate, and vegetation of Latin America have attracted the attention of numerous and often prominent European geographers--e.g., Hettner, Sapper, Sievers, Kühn among an earlier generation, Troll, Kinzl, Maull, Wilhelmy, Czaja, Termer, and Gierloff-Emden more recently--without straying beyond the German-speaking scholars. Then there are Borde, De Martonne, Dresch, Dollfus, Ruellan, and others among the French, and King of South Africa as well as a number of significant Latin American contributors. It is curious indeed that the pioneering work of Bowman[29] in the Central Andes seems to have had no later parallel among U.S. geographers.

The sparseness of the contributions by North Americans to the study of the physical geography of Latin America[30] reflects, of course, the relatively low status of the physical side of our subject, a condition which the recent ecological awakening may be at long last changing. The environment is increasingly being recognized as humanized nature rather than merely as a physical category, a part of the philosophy of man as inhabitant and transformer of the earth. Brunnsweiler, in pointing to the paucity of work on land forms, and climate, most of which has been done by geologists and meteorologists, has urged the importance of a more coherent and practically oriented physical geography that may be relevant to our increasing ecologic and environmental concerns. Only in coastal

---

[28]Wilbur Zelinsky, "Historical Geography of the Negro Population of Latin America," *Journal of Negro History*, Vol. 34 (1949), pp. 153-221.

[29]Isaiah Bowman, *The Andes of Southern Peru* (New York: American Geographical Society, 1916).

[30]Dieter Brunnsweiler, "The Study of the Physical Environment in Latin America," in Barry Lentnek *et al.* (eds.), *Geographic Research on Latin America: Benchmark 1970* (Muncie, Indiana: Ball State University, 1971), pp. 220-231.

studies,[31] thanks in good measure to the program developed by the late R. J. Russell and his colleagues at Louisiana State University, and perhaps in studies of the natural and man-altered vegetation,[32] can we hold our heads at all high. As things now stand, Brunnsweiler notes the average student is likely to get a more meaningful elaboration of the natural factors in the environment of a region from a Madariaga or a Ted Szulc than from geographers.

As man increasingly intervenes in nature, the maintenance of the man-land system in some sort of ecological equilibrium is more and more recognized as our principal challenge and responsibility. In this the involvement of our Latin American colleagues must have the highest priority. One obvious task is identification and analysis in human terms of the extreme and fragile environments that seem more characteristic of Latin America than most other world regions. The problems associated with tropicality and aridity--especially as related to soil fertility, human disease, and water surpluses or shortages--are fundamental concerns over wide areas. Such "short-term phenomena" as earthquakes, volcanic eruptions, landslides, avalanches, hurricanes, floods, drought, and wild fires have also had serious long-term consequences here for man. But there have been few geographic analyses or even case studies of such phenomena[33] and none of the type of hazard research that has come to be so strongly identified with geography at Clark, Chicago, and Toronto.

One of the more encouraging recent developments within geography has been the revitalization of the sub-field of biogeography. Much of this work, and more especially the history of vegetation modification through human activity, has centered in tropical America, where one tends to become especially sensitized to the importance of vegetation as the essential geographic factor that more than anything else defines and gives character to the visible landscape.

---

[31] Norbert Psuty, "Contribution to the Coastal Morphology of Latin America," in Lentnek _et al._ (eds.), _op. cit._ (1971), pp. 250-264.

[32] For bibliographies see James J. Parsons, "Ecological Problems and Approaches in Latin American Geography," in Lentnek _et al._ (eds.), _op. cit._ (1971), pp. 13-32. David de Laubenfels, "The Vegetation Formations of Latin America," _Revista Geográfica_ (Rio de Janeiro), Vol. 72 (1970), pp. 97-138.

[33] But see Hilgard O'R. Sternberg, "Floods and Landslides in the Paraíba Valley, December 1948--Influence of Destructive Exploitation on the Land," _Comptes Rendu_, International Geographical Congress, Lisbon, 1949, Vol. 3 (1951), pp. 335-364; John D. Rees, "Paricutín Revisited: A Review of Man's Attempts To Adapt to Ecological Changes Resulting from Volcanic Catastrophe," _Geoforum_, Vol. 4 (1970), pp. 7-25.

The relationship of plant cover to human activity seems much closer here than in temperate lands so that for the field geographer the ability to make proper identifications of growing things and to recognize indications of disturbance becomes of considerable importance.

Vegetation and vegetation change provide some of the most sensitive indicators of ecological potential and environmental deterioration and it is here that some of the most important contributions by geographers have been made.[34] In numerous explicitly regional studies, too, special attention has been given to vegetation and its modification and use by man.[35] Maps of ecological potential, drawn on the basis of a predictive climatic model created by Leslie R. Holdridge, have been published for almost all of the Spanish-speaking countries from Guatemala to Peru using a terminology relating to plant formations. Perhaps the most sustained efforts have been focused on the vexing question of tropical savannas, including a series of some 16 volumes published by the McGill Savanna Research Project. Geographers have been giving attention to the pampas question,[36] to strand vegetation,[37] and to problems of high mountain geo-

---

[34] Leo Waibel, "Vegetation and Land Use in the Planalto Central of Brazil," Geographical Review, Vol. 38 (1948), pp. 529-554; Joseph A. Tosi, Jr., Zonas de vida natural en el Perú: memoria explicativa sobre el mapa ecológico del Perú (Lima: Inst. Interamericana de Ciencias Agrícolas de la OEA, Zona Andina, 1960); Joseph A. Tosi, Jr. and Robert Voertman, "Some Environmental Factors in the Economic Development of the Tropics," Economic Geography, Vol. 40 (1964), pp. 189-205; Philip Wagner, "Natural and Artificial Zonation in a Vegetation Cover: Chiapas, Mexico," Geographical Review, Vol. 52 (1958), pp. 253-274; Theo Hills, "Savannas: A Review of a Major Research Problem in Tropical Geography," Canadian Geographer, Vol. 9 (1965), pp. 216-228; Leslie R. Holdridge, Life Zone Ecology (with photographic supplement) (San Jose, Costa Rica: Pacific Science Center, 1967); Forest Environments in Tropical Life Zones: A Pilot Study (Oxford: Pergamon Press, 1971); B. L. Gordon, Anthropogeography and Rain Forest Ecology in Bocas del Toro Province, Panama (Department of Geography, University of California, Berkeley [ONR Report], 1969).

[35] B. L. Gordon, "Human Geography and Ecology of the Sinú Country of Colombia," Ibero-Americana, No. 39 (Berkeley and Los Angeles, 1957); Robert C. West, The Pacific Lowlands of Colombia: A Negroid Area of the American Tropics (Baton Rouge: Louisiana State University Press, 1957); R. West, N. Psuty, and B. Thom, The Tobasco Lowlands of Southeastern Mexico (Baton Rouge: Louisiana State University Coastal Studies Institute, 1968).

[36] Oscar Schmieder, "The Pampa--A Natural or Culturally Induced Grassland?" University of California Publications in Geography, Vol. 2 (1927), pp. 255-270; "Alteration of the Argentine Pampa in the Colonial Period," ibid., pp. 303-321; H. Walter, "Das Pampaproblems in vergleichend ökologischen Betrachtung und seine Lösung," Erdkunde, Vol. 21 (1967), pp. 181-202, with rejoinder by C. Troll and W. Lauer, Vol. 22 (1968), pp. 152-158.

[37] Jonathan Sauer, Reconnaissance of Seashore Vegetation along the Mexi-

ecology;[38] the deserts and tropical forest have been more largely left to the botanists and ecologists. All of these major associations have been modified importantly by human activity, but it is the relentless attack on the tropical rain forest that recently has been the object of special concern.[39]

The Impact of Man

Plant succession in disturbed situations such as abandoned agricultural plots provides important clues to the process of vegetational change[40] and this often involves changes in the soil and micro-climate. Weedy immigrant species usually introduced through human agency, have contributed largely to the reconstitution of secondary formations throughout the American tropics[41] as well as frequently being the cause of abandonment of agricultural plots.[42] Selective protection of useful items in the second-growth (rastrojo) has favored certain palms, leguminous trees, chicle, Brosimum, pitahaya, calabash, and a host of other species in many parts of tropical America. Elsewhere exotic invaders, such as cashews, oranges, limes, and mangoes have established themselves so successfully they may be mistaken for natural elements of the vegetation. One

---

can Gulf Coast (Baton Rouge: Louisiana State University Coastal Studies Institute, 1968); John Vann, "Landform-Vegetation Relationship in the Atrato Delta," Annals of the Association of American Geographers, Vol. 49 (1958), pp. 345-360; Robert C. West, "Mangrove Swamps of the Pacific Coast of Colombia," Annals of the Association of American Geographers, Vol. 46 (1956), pp. 98-121.

[38] See especially the papers from the Mexico City Symposium on Geo-ecology of the mountainous regions of tropical America, Colloquium Geographicum (Bonn), Vol. 9 (1968); also Carl Troll, "Die tropischen Gebirge," Bonner Geographische Abhandlungen, Vol. 25 (1959).

[39] For bibliography see A. C. Gômez-Plata, C. Vázquez-Yanes, and S. Guevara, "The Tropical Rain Forest: A Non-Renewable Resource," Science, Vol. 177 (1972), pp. 762-765; also William M. Denevan, "Development and the Imminent Demise of the Amazon Rain Forest," unpublished paper presented at the 1971 CLAG Conference, Syracuse, N.Y.

[40] Gerardo Budowski, "Forest Succession in Tropical Lowlands," Turrialba, Vol. 6 (1963), pp. 23-33.

[41] E.g., Carl Johannessen, "Man's Role in the Distribution of the Coroza Palm (Orbignya spp.)," Yearbook of the Association of Pacific Coast Geographers, Vol. 19 (1957), pp. 29-33; Herbert Eder, "Palms and Man in Coastal Oaxaca," ibid., Vol. 32 (1970), pp. 41-58.

[42] E.g., M. C. Kellman and C. D. Adams, "Milpa Weeds of the Cayo District, Belize, British Honduras," Canadian Geographer, Vol. 14 (1970), pp. 323-343.

of the best examples of this replacement of native by exotic elements is the gradual invasion of African grasses in tropical America, often with the aid of fire or grazing.[43] This establishment is in part on old savanna surfaces but more largely on recently cleared tracts of tropical forest where, after two or three years of cropping, the land has been converted to pasture. In terms of area covered we may be dealing here with one of the greatest ecological invasions of history. Its consequences are uncertain. In the short run, at least it would seem to offer not only superior protein content but resistance to fire and overgrazing. But the durability of such a tropical "grass economy" is questionable. A similar replacement of native grasses by introduced species, from the Mediterranean region, is occurring in middle Chile. In Argentina species from the more humid mid-latitudes of Europe are involved.

It has been shown that grazing animals replaced man in the tropical lowlands of Mexico with the decline of Indian populations in the 16th century, thus promoting the invasion of adventive woody species.[44] Overgrazing may also have so reduced fuel supplies in some areas that fires were eliminated as an ecologic factor[45] with a resultant advantage to unpalatable woody species, especially those adapted to drought. Pines, widely distributed in Middle America, have attracted special attention as successional species adapted to fire, although fire may kill seedlings during the early stages of growth.[46] Ecological assemblages, although not climax or in equilibrium, are thus being established and maintained.

The appropriation of wild plants for use by man has been a recurrent theme in the literature of cultural biogeography. However, animal geography has remained, like ecological plant geography, largely the domain of the professional biologists, although man's use of wild game and fish has tended increasingly to come within the scope of the geographers' concern.[47]

---

[43] James J. Parsons, "The Spread of African Pasture Grasses to the American Tropics," Journal of Range Management, Vol. 25 (1972), pp. 12-17.

[44] Lesley B. Simpson, "The Exploitation of Land in Central Mexico in the Sixteenth Century," Ibero-Americana, No. 36 (Berkeley and Los Angeles, 1952).

[45] Carl Johannessen, "The Savannas of Interior Honduras," Ibero-Americana, No. 46 (Berkeley and Los Angeles, 1963).

[46] James J. Parsons, "The Miskito Pine Savanna of Nicaragua and Honduras," Annals of the Association of American Geographers, Vol. 45 (1955), pp. 36-63; William M. Denevan, "The Upland Pine Forests of Nicaragua," University of California Publications in Geography, Vol. 12 (1961), pp. 251-320.

[47] Charles F. Bennett, "The Bayano Cuna Indians, Panama: An Ecological

One of the more important geographical aspects of Latin America is its predominant Tropicality. Most of its land mass lies within 30° of the Equator. From Mazatlán to the Norte Chico and from Tampico to São Paulo seasonal contrasts in temperature are minimal; a full two-thirds of the area is wet enough to support tropical or sub-tropical forest associations. In such environments, as Gourou, Pendleton[48] and others have pointed out, problems of soil fertility, soil exhaustion, human disease, and dietary deficiency take on a special character quite unfamiliar to mid-latitude observers. The familiar struggle between man and nature seems often weighted in favor of the latter, and the delicately balanced ecosystems may be subject to serious, even irreversible alteration, unless handled with the utmost care. Land abandonment among milpa farmers has not been due solely to soil exhaustion. Grass and weed invasion, insect infestations, and cultural tradition may prove to be at least equally important factors. Yet shortened fallow periods caused by population pressures and the local unavailability of land may lead to impoverishment of second growth and eventually to savanna conditions and ecological disaster.[49] A real controversy rages over the extent to which the tropical lowlands of the New World, and more especially the leached upland surfaces of Amazonia, may be amenable to permanent occupance by sedentary farmers, even if a system of mixed farming with wet rice and tree crops, or a livestock economy, is adopted.[50]

The impact of man's activities on the physical environment of South

---

Study of Livelihood and Diet," Annals of the Association of American Geographers, Vol. 52 (1962), pp. 32-50; Bernard Nietschmann, "Hunting and Fishing Focus among the Miskito Indians, Eastern Nicaragua," Human Ecology, Vol. 1 (1972), pp. 41-67; James J. Parsons, The Green Turtle and Man (Gainesville: University of Florida Press, 1962); David Henderson, Men and Whales at Scammon's Lagoon (Los Angeles: Dawson's Book Shop, 1972).

[48] Robert L. Pendleton, "The Place of Tropical Soils in Feeding the World," Smithsonian Institution Annual Report, 1955, pp. 441-458; Pierre Gourou, The Tropical World: Its Social and Economic Condition, 4th ed. (New York: Longsmans, Green, 1966).

[49] Gerardo Budowski, "Tropical Savannas: A Sequence of Forest Felling and Repeated Burning," Turrialba, Vol. 6 (1956), pp. 23-33; "La influencia humana en la vegetación natural de montañas tropicales americanas," Colloquium Geographicum (Bonn), Vol. 9 (1968), pp. 157-162.

[50] Pierre Gourou, "Observaçaões na Amazonia," Revista Brasiliera de Geografía, Vol. 11 (1949), pp. 355-408; Vol. 12 (1950), pp. 171-250; Betty Meggers, Amazonia: Man and Culture in a Counterfeit Paradise (Chicago: Aldine, 1971).

America has been catalogued by Sternberg.[51] It is an impressive listing, and it has many roots deep in human history. Man has been in Latin America for a long time, longer than any but a Sauer dared to imagine only a few years ago. The evidence for Pleistocene occupation of the New World is irrefutable. Man was well established in South America at a time when the Pleistocene ice sheet still covered much of the land. Between 7000 and 1500 B.C. most of the major plant foods had been domesticated and a form of settled life achieved. Population estimates for pre-Columbian America have been steadily going up, as archaeological and archival data are increasingly brought to bear on the problem. It is now believed by best authority to have exceeded 100 million, a figure that has been attained again only in the past decades. The drastic decline in native populations following the Conquest did not significantly reduce the pressures on the environment. The introduction of European livestock, European crops and European plows brought an acceleration of an already serious soil erosion in many areas, a process that is still going on. Geographical analyses of such erosion, together with the effectiveness of terracing and other devices designed to alleviate it, largely remain to be done. They should be vastly facilitated by use of aerial photography.

Vogt has gone further than anyone in placing the erosion problem in clear perspective,[52] coupling it with continuing and largely uncontrolled population growth. He has specifically urged geographical case studies of the history of land use and productivity of specific areas, especially in Mexico and Central America. More recently conferences sponsored by IUCN and UNESCO and the IGU regional conference in Mexico City have further focused attention on related environmental issues. Sauer has repeatedly pointed to the role of destructive exploitation in the non-western world, arguing that man has long been the ecologic dominant[53] and that true "natural vegetation" probably no longer exists anywhere in Tropical America.

---

[51] Hilgard O'R. Sternberg, "Man and Environmental Change in South America," in E. J. Fittkau et al. (eds.), Biogeography and Ecology in South America (The Hague, 1968), Vol. 1, pp. 413-445.

[52] William Vogt, The Population of Venezuela and Its Natural Resources (Washington, D.C.: Pan American Union, 1946); The Road to Survival (New York, 1948); Comments on a Brief Reconnaissance of Resource Use, Progress, and Conservation Needs in Some Latin American Countries (Washington, D.C.: Conservation Foundation, 1963).

[53] Carl Sauer, "Man in the Ecology of Tropical America," Proceedings, Ninth Pacific Science Congress, Bangkok, 1957, Vol. 20 (1958), pp. 104-110.

Given the latent energy of the continuing population explosion it is clear that Latin America, with a still vigorous three per cent annual increase, is faced with providing for a vastly increased number of mouths, and at higher dietary standards, in the next decades.[54] The continent is crying out for regional land resource and land use evaluations such as Preston James has been urging for years. Now, with the availability of radar and satellite imagery, this seems at last within our grasp. It seems likely that intensification of production in already settled areas holds more promise than the continued forest destruction associated with the opening of new lands which, after a few crops of maize or manioc, most commonly are converted to low intensity pasture use. As the list of endangered habitats and endangered species is extended--and whole populations of Amazonian Indians are threatened with extinction by the mania to "conquer" the no longer "limitless" selva--the move for nature protection and the establishment of national parks or preserves is beginning to take hold. Attention of Latin American governments to such problems has been heightened by the United Nations Conference on the Human Environment held at Stockholm in 1972, but the effectiveness of any measures that may be taken to this end will depend on a wholesale revision of attitudes that is not likely to be quickly or easily achieved.

Contemporary Human Geography

Descriptive accounts of contemporary systems of production and transportation, of human migration and settlement, or rural and urban landscapes--more commonly in the form of journal articles than in-depth monographs--comprise much of the writings of North American geographers on Latin America. They offer information and insight of considerable intrinsic interest, often more pedagogic than scholarly, and are not without utility to planners or politicians. They are perhaps more in the tradition of the geographer in foreign lands as a journalist or explorer, recording the human scene in its geographical context, with concern for cultural particularisms and ecologic consequences, for the generation of hypotheses rather than their proof.

---

[54] Wilbur Zelinsky, "The Race between Population and Resource Development in Central America and the West Indies: A Review of Current Knowledge and Speculation," in W. Zelinsky, L. A. Kosiński, and R. W. Prothero (eds.), Geography and a Crowding World (New York: Oxford University Press, 1970), pp. 511-534; and Gerhard Sandner, "Population Pressure upon Resources in Costa Rica," ibid., pp. 535-555.

A prime motivation behind the recent organization of the Conference of Latin Americanist geographers seems to have been the desire of some for an increased involvement in the contemporary issues of development and modernization, and for the focusing of more work specifically on the overriding social, economic, and ecological problems facing the continent by use of a more sophisticated theory and modern quantitative methods. The second CLAG conference at Syracuse was almost exclusively devoted to papers on development and the third, planned for 1973 in Calgary has for its theme "Geographical Analysis for Development Planning." The program chairman of the initial Muncie conference did not hesitate to state his position bluntly:[55]

> Pre-Columbian and other historical studies by geographers are a nicety well worth preserving for a handful of our colleagues and their academic audiences. They should not, however, form the bulk of our literature during the coming crucial decades _if_ we wish our work to be of some service to the tens of million Latin Americans who must, perforce, live through the agonies of modernization during the next generation.

Others urged greater emphasis on theory, on similarities and generalizations rather than differences and uniqueness, along with more emphasis on quantification and more formalized theory. It was warned that regional specialization too often has been a refuge in geography for those "unwilling or unable to contend with the challenges of change." Some proposed that Latin Americanists reorient their priorities towards what have rather glibly been called "the new mainstreams of geographic analysis." Others, less concerned with method than with problem selection, have urged a "more gutty," functional geography.[56]

The geography editor of the biennial _Handbook of Latin American Studies_ (1973), which is the best single source for an overall review of geographical literature on Latin America, has taken vigorous exception to the programmatic exclusiveness represented at Muncie, where no place was found on the program for historical or cultural geography, perhaps the two most vigorous arms of the discipline in the Latin American context. At the 1971 CLAG meetings at Syracuse, too, there was a strong reaction from the more ecologically-oriented participants. There were papers on the destruction of the Amazon forest, on sewage as a pollutant in Central America, on the cost of de-pollution in Brazil,

---

[55] Barry Lentnek _et al._ (eds.), _Geographic Research on Latin America: Benchmark 1970_ (Muncie, Ind.: Ball State University, 1971), p. 161.

[56] Ray F. Watters, "Geography as a Social Science: A More Functional Geography," _East Lakes Geographer_, Vol. 6 (1970), pp. 5-25.

on the social cost of meat export from Central America, and environmental deterioration on the coast of El Salvador. But Charles Bennett likened the representation of "ecology" on a "development panel" as akin to the current need to have minorities represented on establishment committees. To him it is "one of the most astonishing manifestations of our day that so many persons who claim to have focused their attention on man-land relationships should choose to hold the ecological implications of such relationships at arms length, talking about "development" whilst about their heads an ecological storm gathers that threatens the lives of millions." He expressed special dismay that so many geographers who have chosen to concern themselves with the economic problems of Latin America have "copped out" on the issue of population growth and have given their attention to the far less controversial area of "development." He observed that if population growth is not checked, all talk about development could only serve as an exercise in Kafkian rhetoric. But he gave no prescription for the concerned geographer, instead falling back on Sauer's rhetorical question: "Are we the cleverest people of all time or the blindest because we think neither whence nor whither we are bound?"

Such programmatic pleas for a reorientation of geographic research as those emanating from Muncie and Syracuse have as yet been little reflected in the literature. There are still relatively few substantive works in geography directed towards contemporary development problems and the optimizing of investment or location, while the much heralded "new geography," in which the particular case is considered significant chiefly as a source of data to be used for generalizations, might as well not have existed so far as Latin America is concerned.[57]

It is worthwhile asking why this is so, and it is tempting to suggest that the inherent interest of the diverse historical, cultural, and environmental factors as they converge in given Latin American localities, and their recognizably dominant significance, is in good part responsible for the low state of geograph-

---

[57]There are, of course, exceptions. See R. F. Semple and H. L. Gauthier, "Spatial and Temporal Trends in Income Inequalities in Brazil," Geographical Analysis, Vol. 4 (1972), pp. 169-180; Peter Gould and Jack P. Sparks, "The Geographic Context of Human Diets in Southern Guatemala," Geographical Review, Vol. 59 (1969), pp. 58-82; Barry Lentnek, "Economic Transition from Traditional to Commercial Agriculture: The Case of El Llano, Mexico," Annals of the Association of American Geographers, Vol. 59 (1969), pp. 65-84; R. N. Thomas, "The Migration System of Guatemala City: Spatial Inputs," Professional Geographer, Vol. 24 (1972), pp. 105-112.

ical theory and formal spatial analysis in this area. It is rare that anyone with command of Spanish or Portuguese--and unlike Africa, Asia, or the Middle East significant work here is completely precluded without control of the language--has found much appeal or applicability in the more vigorous sort of locational analysis or model-building when confronting the realities of the Latin American scene. The human and environmental conditions are simply too obviously intriguing and significant. At least as challenging as the engineering of social change is the explanation for its absence in some communities, its presence in others.[58] Indeed, the cultural particularism, the dynamism, of such groups as the Antioqueños or the Paulistas, for example, may often be the most relevant single factor in explaining economic development and cultural change. Yet even the behavioral component of problems of such a fundamental nature in human geography have been but weakly developed.

Geographers with a cultural-historical bias have on the whole tended to be a pragmatic group, questioning the optimistic view that Latin America is on the same unilinear track towards economic development that developed Western nations rode to prosperity some time ago. Some have even questioned seriously the appropriateness of our putting our fingers on the balance wheel at all, of encouraging or facilitating a "development" process on the Western model based on diminished diversity and increased consumption, with the accelerating drain on the resource base that accompanies it.

It comes back to the familiar theme of differing, and perhaps increasingly polarized, temperaments. One group looking on the environment through the eyes of the humanist, focuses on the "wonder and curiosity" of the world as given, on man's dependence on nature, and what he is doing to it. For the social scientists, the goal is more explicitly the understanding of society, how man relates to man, often with greater concern for theory and prediction than with pattern or place. While members of either group may be striving for a more functional or "relevant" geography appropriate to the immediate pressures of the day, action-oriented research is in fact more characteristic of the second than the first. Perhaps the area where these two distinct attitudes and approaches converge most closely is in the evaluation and inventory of land re-

---

[58]Ronald C. Scheck, "The Persistence of Cultural Autarchy," in Barry Lentnek *et al.* (eds.), *Geographic Research on Latin America: Benchmark, 1970* (Muncie, Ind.: Ball State University, 1971), pp. 350-355; H. Homer Aschmann, "Indian Societies and Communities in Latin America: An Historical Perspective," *ibid.*, pp. 124-137.

sources and the integration of this information with economic and social possibilities and goals so crucial to the planning process. In this work location theory, central place studies, and the standard techniques of spatial analysis may find a role.

Geographers with the Office of Regional Development of the Organization of American States (OAS) under Kirk Rogers and Wolfram Drewes (now with the World Bank) have been largely responsible for the splendid portfolios of maps covering principal environmental parameters of such areas as the La Plata and Guayas river basins, the Dominican Republic, and Haiti. Several country atlas projects have also been under their direction. The Army Map Service, collaborating with local Latin American governments and using advanced cartographic and photogrammetric techniques, has given us detailed map coverage of many parts of Latin America superior to that which exists for even the United States. Earlier the Puerto Rico Land Classification Survey, conducted under the joint direction of U. S. and Puerto Rican geographers, provided another example of such collaboration. Now further new advances in technology, such as represented by the RADAM (Radar in Amazonia) and the ERTS (Earth Resources Technology Satellite) projects promise further undreamed of opportunities which we are inevitably compelled to take advantage of. In the end geographers interested in the field of development planning, if they are to serve up anything but warmed over economics and sociology, must return to the field and address themselves to the concept of the region and regional analysis.[59] Perhaps it is true that nothing less than a systems analysis (whatever that is!) of the whole complex of natural and cultural processes can yield the full background for decision making. We must recognize, of course, that planning is a political process, its implication often an economic problem, but it is the failure to recognize fully the ecologic base on which it is dependent that has led to some of its most conspicuously dismal failures.[60]

Land Tenure and the New Colonization Schemes

One of the more pressing problems in Latin America is the uneven distribution of land and the inefficiencies in production that this has often engen-

---

[59] Richard Momsen, "Mapping Spatial Factors for Development Planning," _ibid._, pp. 397-401.

[60] Pierre Monbeig, "Points de vues géographiques sur le sous-développement en Amérique Latin," _Annales de Géographie_, Vol. 76 (1967), pp. 704-713.

dered. Latifundia and minifundia have deep historical and cultural roots, but despite the early classic studies of McBride,[61] few geographers have pursued the subject in any depth. It has been economists, political scientists, and sociologists, as at the Wisconsin Land Tenure Center, who have done most of the work in this field, although often oblivious of the ecological factors involved. Government colonization schemes, on the other hand, have attracted considerable attention from geographers.[62] So has export-based tropical plantation agriculture and other types of large-scale corporate agricultural enterprises, especially when associated with irrigation developments.[63]

The opening of "new lands" in the _tierra caliente_ and the impact of the _vias de penetración_ that have been reaching down from the Central American and Andean highlands into the adjacent lowlands and from the Brazilian coast into the heart of Amazonia has probably been the single most intensively pursued theme within Latin American geography.[64] A recurrent theme in this has been

---

[61] George M. McBride, _The Land Systems of Mexico_ (New York: American Geographical Society, 1923); _Chile: Land and Society_ (New York: American Geographical Society, 1936).

[62] C. W. Minkel, "Problems of Agricultural Colonization and Settlement in Central America," _Revista Geográfica_ (Rio de Janeiro), Vol. 40 (1967), pp. 19-53; Robert Eidt, "Comparative Problems and Techniques in Tropical and Semi-Tropical Pioneer Settlement: Colombia, Peru, and Argentina," _Yearbook, Association of Pacific Coast Geographers_, Vol. 26 (1964), pp. 37-42.

[63] E.g., David A. Henderson, "Arid Lands under Agrarian Reform in Northwest Mexico," _Economic Geography_, Vol. 41 (1965), pp. 300-312; Pierre A. D. Stouse, Jr., "Instability in Tropical Agriculture: The Atlantic Lowlands of Costa Rica," _Economic Geography_, Vol. 47 (1971), pp. 78-97.

[64] E. g., Robert Eidt, "Pioneer Settlement in Eastern Peru," _Annals of the Association of American Geographers_, Vol. 52 (1962), pp. 255-278; _Pioneer Settlement in Northeast Argentina_ (Madison: University of Wisconsin Press, 1971); Norman R. Stewart, "Migration and Settlement in the Peruvian Montaña: The Apurimac Valley," _Geographical Review_, Vol. 55 (1965), pp. 143-157; "Some Problems of Agricultural Colonization in the Andean Oriente," _Professional Geographer_, Vol. 20 (1968), pp. 33-38; Alfonso González, "Problems of Agricultural Development in a Pioneer Region of Southwest Coastal Mexico," _Revista Geográfica_ (Rio de Janeiro), Vol. 64 (1966), pp. 29-52; E. E. Hegen, _Highways into the Upper Amazon Basin: Pioneer Lands in Southern Colombia, Ecuador and Northern Peru_ (Gainesville: University of Florida Press, 1966); Alfred Siemens, "New Agricultural Settlement along Mexico's Candelaria River: Implications of Commitment to Planning and the Ejido," _Interamerican Economic Affairs_, Vol. 20 (1966), pp. 23-39; James J. Parsons, "Antioquia's Corridor to the Sea: An Historical Geography of the Settlement of Urabá," _Ibero-Americana_, No. 49 (Berkeley and Los Angeles, 1967); C. L. Dozier, _Land Development and Colonization in Latin America: Case Studies of Peru, Bolivia, and Mexico_ (New York, 1969).

the contrasts between spontaneous and planned colonization in such areas and the cultural adaptations of highlanders (often Indians) to a drastically different lowland environment. The recent literature on agricultural colonization in Latin America has been well reviewed by the sociologist, T. Lynn Smith.[65] The process of forest clearing and its replacement, chiefly by pasture, is visibly conspicuous and peculiarly adaptable to geographical analysis, as the classic studies by Sandner in Costa Rica and Central America have shown.[66] More recently there has been a spate of excellent German-language research on forest clearing and agricultural settlement along the Andean front. The ecologic effects of this massive alteration of the face of the tropical lowlands by man is but dimly perceived. In some areas archaeologic evidence of formerly large native populations suggest that production systems other than the popular Western monoculture model may offer more efficient use of the land.[67] The frontier philosophy views the conquest of the forest with pride and enthusiasm, much as North American pioneers saw the replacement of woodland by cropland as inevitable and desirable. For Brazil, for example, the Belem-Brazilia and Trans-Amazon highway projects epitomize the national aspirations to open the empty interior to settlement and civilization, although a kind of genecide of the Indian inhabitants of the Amazonian forest may be a by-product of this process. As long as land is viewed as a cheap resource intensification of effort on existing settled lands will be delayed and the destruction of the rain forest ecosystem will continue.

Agricultural colonization often has been by cohesive cultural groups in

---

[65]T. Lynn Smith, "Studies of Colonization and Settlement," Latin American Research Review, Vol. 4 (1969), pp. 93-124.

[66]Gerhard Sandner, "Agrarkolonisation in Costa Rica," Schriften des Geogr. Inst., Univ. Kiel, Vol. 20 (1961), No. 3; "El concepto espacial y los systemas functionales en la colonización espontanea Costarricense," Informe Semestral, Inst. Geogr. Costa Rica, San Jose, 1962; "Die Erschliessung der Karibischen Waldregion im Südlichen Zentralamerika," Die Erde, Vol. 95 (1964), pp. 111-131. Sandner's principal works have been published in Spanish translation by the Instituto Geográfico, San Jose. Another outstanding study of pioneer settlement by a German geographer is W. Brücher, "Die Erschliessung des tropischen Regenwaldes am Ostrand der kolumbianischen Anden," Tübingen Geographische Studien, No. 28 (1968). See also A. Maass, "Entwicklung und perspectiven der wirtschaftlichen Erschliessung des Tropischen Waldlandes von Peru," ibid., No. 31 (1970).

[67]J. C. Dickinson III, "Alternatives to Monoculture in the Humid Tropics of Latin America," Professional Geographer, Vol. 24 (1972), pp. 217-222; B. L. Gordon, Anthropogeography and Rain Forest Ecology in Bocas del Toro Province, Panama (Department of Geography, University of California, Berkeley [ONR Report], 1969).

temperate as well as tropical Latin America and has been the object of our frequent attention.[68] Although rich in historical detail, such studies seldom have been characterized by the innovative research designs associated with work in some other parts of the tropical world.

Urban and Political Geography

The conspicuous neglect of the phenomenon of urbanization by geographers working in Latin America is not readily explainable. As the organization center of human society, exercising powerful influences on the use of surrounding space, and as a geographical phenomenon of interest for its own sake, whether in terms of its origins, morphology and functions, its external linkages or its role as a center of innovation and political power, the Latin American city offers a multitude of significant and unexplored geographical themes. The historian Richard Morse[69] recently has pointed to many of these themes in a remarkable and stimulating review essay. This continuing neglect is the more remarkable in view of the extensive and growing literature on urban origins and the critical role of cities in the rise of civilization, in Mesoamerica and Peru as well as in the Old World. Judged from the published literature there are few North American "urban geographers" with more than a casual commitment to the Latin American city,[70] a remarkable situation in view of the highly developed nature of urban geography generally as a sub-field within our discipline. Among geographers we must turn to Europeans[71] for the major works on city form and

---

[68] E.g., Norman R. Stewart, Japanese Colonization in Eastern Paraguay, NAS-NRC Publication 1490 (Washington, D.C., 1967); Morton Winsberg, "Jewish Agricultural Colonization in Entre Rios, Argentina," American Journal of Economics and Sociology, Vol. 27 (1968), pp. 285-296; Vol. 28 (1969), pp. 179-181, 423-428; Glyn B. Williams, "The Welsh in Patagonia: A Geographic Perspective," Revista Geográfica (Rio de Janeiro), No. 69 (1968), pp. 121-144; Leonard Sawatsky, They Sought a Country: Mennonite Colonization in Mexico (Berkeley and Los Angeles: University of California Press, 1971).

[69] Richard M. Morse, "Trends and Issues in Latin American Urban Research, 1965-70," Latin American Research Review, Vol. 6 (1971), pp. 3-52, 19-75.

[70] E.g., Lawrence Brown, Clyde Browning, Gerald Pyle, Charles Sargent, and David Snyder. Significantly the only paper on "urbanization" at the 1970 CLAG conference at Muncie was presented by an Africanist! But see Charles Sargent, "Toward a Dynamic Model of Urban Morphology," Economic Geography, Vol. 48 (1972), pp. 357-375, using a Buenos Aires case study, for evidence that this neglect of urban geography may be ending.

[71] E.g., E. Gormsen, "Barquisimeto, eine Handelstädt in Venezuela,"

function. The most widely quoted papers on Latin American urbanization by a U.S. geographer, curiously, are those on the origins of the colonial city and grid pattern towns by Dan Stanislawski, whose more recent substantive work has been focused on quite different cultural themes in the Mediterranean area.[72]

Transportation problems, often related to migration flows and economic development potential, are beginning to attract the attention of geographers.[73] Such work may increasingly complement that in the rapidly expanding field of contemporary population research,[74] the subject of a review volume scheduled for publication by the Conference of Latin American Geographers in early 1973 (Population Dynamics of Latin America: A Review and Bibliography, Robert N. Thomas, ed. [East Lansing: CLAG Publications]).

Political geography, too, has received but relatively little attention in

---

Heidelberg Geographische Arbeiten, Vol. 12 (1963); Herbert Wilhelmy, Südamerika im Spiegel seiner Städte (Hamburg, 1952); Gerhard Sandner, Die Hauptstädte Zentralamerikas (Heidelberg, 1969); Yves LeLoup, "Les villes du Minas Geraes," Travaux et Memoires, 25, Institut des Hautes Études de l'Amerique Latine (Paris, 1970); James M. Houston, "The Foundation of Colonial Towns in Hispanic-America," in E. P. Beckinsale and J. M. Houston (eds.), Urbanization and Its Problems (Oxford: Oxford University Press, 1968), pp. 352-391. The numerous publications of the Institut des Hautes Etudes de l'Amerique Latine, Paris, include many titles by geographers concerned with urban themes. Latin American Urban Research, an important new annual published by the Center for Latin American Studies at the University of Florida, Gainesville, carries an exhaustive international bibliography on urban and urban-related topics and provides a continuing forum for Latin American urban scholars. There is an impressive amount of work going on, but curiously little of it is by geographers.

[72] Dan Stanislawski, "Early Spanish Town Planning in the New World," Geographical Review, Vol. 37 (1947), pp. 94-105; The Anatomy of Eleven Towns in Michoacán (University of Texas, Austin, Latin American Studies No. 10, 1950); "The Origin and Spread of the Grid Pattern Town," Geographical Review, Vol. 36 (1946), pp. 105-120.

[73] Reviewed by Howard Gauthier, "Geography, Transportation and Regional Development," in Barry Lentnek et al. (eds.), Geographic Research on Latin America: Benchmark, 1970 (Muncie, Ind.: Ball State University, 1971), pp. 333-342.

[74] See P. D. Simkins, "Contributions by Geographers to Latin American Population Studies," ibid., pp. 72-85; Pierre A. D. Stouse, Jr., "Settlement Geography in Latin America," ibid., pp. 95-103; R. N. Thomas, "Internal Migration in Latin America: An Analysis of Recent Literature," ibid., pp. 104-118; and G. T. Trewartha, The Less Developed Realm: A Geography of Its Population (New York: Wiley, 1972).

Latin America except for studies of the almost unending boundary disputes[75] that have characterized the continent's last 150 years. Yet political organization of space is of extraordinary importance in the explanation of Latin American economic patterns, not the least in historical terms. Conflicts over the extension of territorial seas (the 200-mile limit), the controversies over the Malvinas (Falklands), the Panama Canal, the Antarctic, and the division of international waters[76] are but a few of the geographical issues that continue to becloud parts of the continent. The settlement of the El Chamizal dispute between the US and Mexico suggests that there are viable answers, but they often come with difficulty. At another level the geographical aspects and consequences of foreign capital investment and its treatment, and the international agreements relating to commodities such as coffee and petroleum, for example, have especial importance for the Latin American economy. So too do the regional treatises and quasi-economic political units, such as the Central American Common Market and the Latin American Free Trade Association which are seeking to improve Latin America's terms of trade with the rest of the world. Methods of economic regionalization and the notion of "growth poles," regional income, investment disparities, primate city dominance, and the conflicts and even separatist movements that these have engendered, are attracting increasing attention in the social science literature.[77] But the geographers' input tends to lag. A colloquium of the IGU Commission on Regional Aspects of Economic Develop-

---

[75]E.g., R. S. Platt, "Conflicting Territorial Claims in the Upper Amazon," in C. C. Colby (ed.), Geographic Aspects of International Relations (Chicago: University of Chicago Press, 1938), pp. 241-276; J. V. Fifer, "Bolivia's Boundary with Brazil: A Century of Evolution," Geographical Journal, Vol. 132 (1966), pp. 360-372.

[76]M. I. Glassner, "The Rio Lauca: Dispute over an International River," Geographical Review, Vol. 60 (1970), pp. 192-207.

[77]See especially John Friedmann, Regional Development Policy: A Case Study of Venezuela (Cambridge, Mass., 1966), a seminal work on the regionalization of socio-economic space that has attracted much attention from more theoretically oriented geographers and planners, but not from geographers working in Latin America. For a critique of Friedmann, see Roland Artle, Urbanization and Economic Growth in Venezuela, Working Paper 135, Institute for Urban and Regional Development, University of California, Berkeley, 1970. See also Brian J. L. Berry, "Relationships between Regional Development and the Urban System: The Case of Chile," Tijdschrift voor Economische en Sociale Geografie, Vol. 60 (1969), pp. 283-307, for an example of a North American economic geographer's application of modern geographical method in a Latin American context.

ment held in Brazil in 1971 on "Regional Inequalities of Development," gave special emphasis to Latin America. Participants included five North American geographers as well as economists, sociologists, and planners.

Regional Geography

The more ambitious regional works have been a major part of the European geographical tradition, but hardly characteristic of North Americans. Of the nearly 50 representative regional monographs cited in my 1964 review of geographical work on Latin America, only eight were by North Americans.[78] The more thematic or historical studies[79] were not considered. Among the more recent outstanding regional syntheses with which I am familiar, only one has been by a North American author[80] although the inclusion of explicitly his-

[78]They were G. M. McBride, Chile: Land and Society (New York: American Geographical Society, 1936); Henry Sterling (ed.), Problemas económicas y sociales de los Andes venezolanos (Caracas: Consejo de Bienestar Rural, 1955); Robert C. West, The Pacific Lowlands of Colombia; A Negroid Area of the American Tropics (Baton Rouge: Louisiana State University Press, 1957); Charles Alexander, "The Geography of Margarita Island, Venezuela," University of California Publications in Geography, Vol. 12 (1958), pp. 85-192; Raymond E. Crist, Etude géographique des Llanos du Venezuela occidental (Grenoble, 1937); The Cauca Valley, Colombia: Land Tenure and Land Use (Baltimore, 1952); Philip Wagner, "Nicoya: A Cultural Geography," University of California Publications in Geography, Vol. 12 (1958), pp. 195-250; Donald Brand, Coalcomán and Matines del Oro, an ex-distrito of Michoacán, Mexico (The Hague, 1960).

[79]E.g., H. H. Aschmann, "The Central Desert of Baja California: Demography and Ecology," Ibero-Americana, No. 42 (1959); C. O. Sauer, "Colima of New Spain in the Sixteenth Century," ibid., No. 29 (1948); B. L. Gordon, "Human Geography and Ecology of the Sinú Country, Colombia," ibid., No. 39 (1957); James J. Parsons, "Antioqueño Colonization in Western Colombia," ibid., No. 32 (1949).

[80]R. C. West, N. Psuty, and B. Thom, The Tabasco Lowlands of Southeastern Mexico (Baton Rouge: Louisiana State Coastal Studies Institute, 1969). This is a remarkably comprehensive and insightful study, regional geography at its best and worthy of emulation. But it is the Europeans and the Latin Americans themselves who are doing the regional geographies. See, for example, P. Cunill, L'Amerique Andine (Paris, 1966); C. Collin Delavaud, Les régions cotieres du Pérou septentrional (Lima, 1968); O. Dollfus, "Le Pérou: Introduction géographiques à l'étude du développement," Travaux et Memoires, 22, Institut des Hautes Etudes de l'Amerique Latine, Paris; H. Gierloff-Emden, Mexico: Eine Länderkunde (Berlin, 1970); W. Weischet, Chile, Seine Länderkundliche Individualität und Struktur, Wissenschaftliche Landerkunden, Vol. 2/3 (1970); D. Browning, El Salvador: Landscape and Society (Oxford: Oxford University Press, 1971); H. Blume, Die Westindischen Inseln (Braunsweig, 1968); J. V. Fifer, Bolivia: Land, Location and Politics since 1825 (Cambridge: Cambridge University Press, 1972).

torical geographies would somethat bolster the record.[81]

The textbook surveys of the continent remain dominated by Preston James' Latin America, now in its fourth revised edition (1969). It is a detailed, country-by-country account, encyclopedic in content but with significant original interpretations. West and Augelli's Middle America (1966) is essentially cultural-historical in orientation, in the tradition of the earlier syntheses by Oscar Schmieder,[82] but with much fresh material and new insights. J. P. Cole's Latin America, an Economic and Social Geography (1965) is the only topically oriented geography addressed to contemporary political, social, and economic concerns with special attention to problems of modernization and economic development supported by some relatively simple quantitative analysis and models. Robert Platt's Latin America: Countrysides and Regions (1943) represented an experiment in micro-regional sampling, an interesting outgrowth of a school of geography that flourished in the U.S. Middle West in the 1930's. Jean Dorst's, South America and Central America (1967), handsomely illustrated and translated from the French, emphasizes the natural history and ecological problems associated with continuing population growth. Latin America: Geographical Perspectives, edited by Harold Blakemore and C. T. Smith (1971) contains original country or regional essays by several different British geographers.

Conclusion

There are probably at least 300 professional geographers in North America with an active area concern in Latin America. Probably fewer than 30 of these have accounted for half or more of the published English-language literature. Some 10 per cent of all geography Ph. D. dissertations in the decade 1960-70 were on Latin American topics and the figure is probably rising. The Professional Geographer's latest interim listing of completed and in-progress theses included 9 localized in Mexico, 12 in Central America, and 5 in South America. The rise of interest in Central America is worth special note, per-

---

[81] E.g., William M. Denevan, "An Aboriginal Cultural Geography of the Llanod de Mojos of Bolivia," Ibero-Americana, No. 48 (1966); C. O. Sauer, The Early Spanish Main (Berkeley and Los Angeles: University of California Press, 1966); J. J. Parsons, "Antioquia's Corridor to the Sea: An Historical Geography of the Settlement of Urabá," Ibero-Americana, No. 49 (1968).

[82] O. Schmieder, Länderkunde Südamerikas (Leipzig and Vienna, 1932); Länderkunde Mittelamerikas (Leipzig and Vienna, 1934); Länderkunde der Neue Welt, I., Mittel und Südamerika (Stuttgart, 1962).

haps a reflection of both improved highway access and the existence of several organized research and instructional programs there (e.g., the Organization for Tropical Studies in Costa Rica). Our publication record is less impressive. There have probably been as many monographs on Latin American topics published by French and German geographers in the last decade as the combined English language output. This is indicative less of the relative input of research effort than our lack of publication outlets and policy. For virtually all German and French dissertations publication is mandatory. Of the English Ph. D. 's it seems likely that fewer than one in ten appear in print although the majority are probably abstracted in part in journal articles. The charitable view would be that this is no reflection on their quality. There has been a recent and significant revival of Latin American interest among British geographers, in part a consequence of the Parry Committee report which gave a strong push to Latin American Studies in the United Kingdom. The cadre of active workers, all newly involved in the area, includes Browning, Crossley, Donkin, Fifer, Fox, Harris, Preston, Robinson, Stoddart, C. T. Smith, and Houston. Even remote Australasia has at least two major contributors (Butland, Watters). The work of this active group, like that of North Americans, has been to date strongly cultural and historical in its orientation, and based on extended periods of field study.

The establishment of CLAG, with more than 200 dues-paying members, holds the promise of a revitalization of Latin American work among North American geographers. In some quarters there is an almost missionary zeal to promote the use of new methods and techniques to this end. The kind of understanding we seek is not, of course, exclusively subject to examination either by analytic method or by humanistic discourse. Our task here, as in the other world areas, is to bridge the gap between our culture and another. In the context of area studies it is to seek understanding of the interaction of mankind and his environment in a particular regional context, some part of Latin America with which we are more than commonly familiar, taking account of similarities and uniqueness alike. The essential interests of geography are clear enough--the diversity of the earth, the patterns of diversity and repetition, and often of how things came to be and the process of their alteration.[83] The danger is that we may spend ourselves in exhortation, telling each other what

---

[83]E.g., Carl O. Sauer, "The Quality of Geography," California Geographer, Vol. 9 (1970), pp. 5-9.

we ought to be doing and how to do it. Better by far that we get on with the task, each in his own, idiosyncratic way.

CHAPTER 3

THE CARIBBEAN REGION

David Lowenthal
University College, London

Even as vague a charge as the one tendered to us regionalists calls for some accounting of the choice of regions in this discussion. Unless we are content with the assertion that "area specialists study these areas," our explanations would doubtless tabulate each region's special characteristics of land, people, culture, institutions, ecology, and history. But because the aspects highlighted would differ with each area, the resulting regional vignettes would never be strictly comparable. More specifically germane to our aims might be specific comparisons of salient regional characteristics within the same framework. Whether or not area studies disclose inherent regional truths, our composite selections should at least suggest something about the general nature of regionality.

Bases for selecting regions to study come readily to mind. Spatial and territorial extent is one set of rationales--numbers of inhabitants, size and contiguity of land and water surfaces. A second set concerns homogeneity and differentiation: the extent to which lands and peoples within the chosen confines exhibit uniform or heterogeneous traits. A third set concerns structural and functional interaction: the extent to which parts of the whole region are culturally and socially interlinked, administratively and economically interdependent, and the persistence of these connections through time. Awareness comprises a fourth set: degrees of consciousness of and self-consciousness about mutual identities, tribal or territorial, regional or national or global, and the trials and triumphs that such awareness engenders. Self-awareness in "Japan" or "Russia," for example, is obviously unlike that in "Africa" or "Latin America," not only because the former are nation-states and the latter continental congeries but also because their social and cultural identities stem from ethnic and imperial histories dissimilar in content, context, and corporate purpose.

Choices for regional study may also be determined by operational or pragmatic considerations, stemming from the selectors' own history, location, culture, and self-awareness. It is worth underscoring the obvious; the only parts of the world not included in these discussions are the United States itself and those parts of the Western world--Western Europe, Canada, perhaps Australia and New Zealand, or are these considered in the Pacific?--so connected by background or so familiar in culture that they do not count as "foreign," much as the British scholar Banister Fletcher differentiated "architecture or historical principles" from all those egregious and outlandish "nonhistorical" structures east of the Bosphorus, south of Sicily, and west of the Straits of Gibraltar.[1]

As a Caribbeanist I am acutely aware that American national perspectives influence our regional selections. Would an area studies conference sited in England or France, indeed anywhere in the Old World, be likely to specify the Caribbean for analysis on the same scale as South or Southeast Asia or Africa? Yet in our joint enterprise here in Chicago I am allotted just as much time and just as many pages as Messrs. Fisher, Murphey, Soja, and Sopher. Contiguity to, and relationships with, the United States clearly affect the regional coverage we are asked to survey.

The Caribbean in the Context of Other Areas

Beyond the pragmatic American considerations I have just alluded to, objective critera would hardly warrant including the Caribbean in this conference. Let me specify precisely why this is so, in terms of several of the points enumerated above.

Numbers and size

The Caribbean is by far the smallest and least consequential area of the ten in both population and land surface. The people of the non-Hispanic West Indies (including mainland Caribbean areas of like culture) number only about 12 million; and even if Jim Parsons cedes me Cuba, Puerto Rico, and the Dominican Republic from Latin America the total would not reach 30 million. Most of the other realms range upward from 200 million, and even Japan or the Pacific exceed 100 million. Thus the Caribbean is the least populous of all areas by a whole order of magnitude. This is enormously consequential for Caribbean as opposed to other regional interpretations, as I mean to show further on.

---

[1]Banister Fletcher, A History of Architecture on the Comparative Method, 15th ed. (New York: Charles Scribner's Sons, 1950), p. 888.

The same disparity holds, with little modification, for territorial extent. The land area of all the Caribbean islands together is not one tenth that of any other realm. Even if the largely uninhabited continental Guianas are included, the superficies of the Caribbean approach only Japan and the Pacific in magnitude. So miniscule is the Caribbean by contrast with other Western Hemisphere regions that it often becomes a source of error and embarrassment. Latin Americanists, for example, are prone to include the Caribbean as part of their own bailiwick substantively but not cartographically, or vice versa. Thus CLAG's Geographical Research on Latin America: Benchmark 1970 has two maps of South America, one on the back cover, but nothing at all for Middle America, although fully one-third of the works referred to concern Central America and the Antilles.[2] Even geographers engaged in delimiting regional boundaries display a pervasive continental bias.

By contrast with most other regional studies areas, therefore, the Caribbean is really a microcosm. But it is not only a microcosm: it is a set of micro-microcosms, isolated from one another by empty stretches of water. Other than the Pacific, the Caribbean is the only essentially dispersed (or "marine," in Brookfield's term) region to be discussed here. Japan and Southeast Asia are, to be sure, partly noncontiguous island groups. But each includes a relatively small number of discrete land masses, with much greater populations than most Caribbean or Pacific islands. In Japan and in Southeast Asia most people are not divided by the sea. In the Caribbean and the Pacific insularity is a cardinal fact of everyday life. Pacific islanders constantly seek to overcome their isolation; Caribbean islanders talk of regional trade and dream of federation but prefer to live apart.

Duration

The Caribbean is not only the smallest, it is historically the newest of all our regions by any standard of indigenization or cultural connection. Natives of other areas trace their traditions back for millennia, their folk memories nourish present-day sociocultural patterns and local self-awareness. In the Caribbean there are no natives; the American Indians are virtually extinct, and existing peoples, cultures, and social institutions are all post-Columbian. Because they are strictly nonindigenous, As V. S. Naipaul says, "these Caribbean

---

[2]Barry Lentnek, Robert L. Carmin, and Tom L. Martinson (eds.), Geographic Research on Latin America: Benchmark 1970 (Muncie, Indiana: Ball State University, 1971).

territories are not like those in Africa or Asia, with their own internal reverences, that have been returned to themselves after a period of colonial rule."[3] Instead, whether in fief or in freedom, the Caribbean mainly reflects other cultures of which it is both recently and still derivative.

The United States and Canada are, to be sure, little less devoid of aborigines and they likewise lack a distinctive native culture. Caribbean and Anglo-American similarities, by contrast with the rest of the world, are manifold. But there are also fundamental differences: Anglo-America is proud of itself, whereas the Caribbean does not yet know itself, and its people reject much of what they do believe about themselves. Conscious that they have no unique high culture of their own, but only elements derived from other cultures, West Indians take more pride in what they conceive to be their original European, and recently Indian, or African, traits, than in what they feel to be the unsatisfactory melange of these elements.

Pragmatic considerations

The Caribbean is physically closer to the United States than any other "foreign" area, excluding peripheral parts of Mexico. In the light of its minuscule population and tiny land mass, only its American proximity, as I have suggested, can explain our inclusion of the Caribbean at all. United States links with the Caribbean are historically and culturally closer than with any other area surveyed here. Americans and West Indians share legacies of colonialism, slavery, and a European background that has shaped us to the virtual exclusion of indigenous elements, and to a far greater extent than other foreign elements. Though ethnically Afro-Asian, the Caribbean is West European in somewhat the same fashion as Anglo-America; both realms are, in this sense, wholly unlike Latin America.

In one or two particular criteria, then, the Caribbean resembles Latin America or the Pacific. But in most major respects, it departs from the territorial, historical, social, cultural, and political patterns of the other nine regions.

Geographical Research in the Caribbean

Unique as a region, the Caribbean is likewise unique in the way geographers have generally studied it. To geographers in the academic mainstream,

---

[3] V. S. Naipaul, "Power?" (New York Review of Books, Sept. 3, 1970), in his The Overcrowded Barracoon and Other Articles, pp. 246-254, ref. to p. 254.

Caribbean practitioners of the discipline sometimes seem extraordinarily aberrant if not positively ungeographical. For they have ignored or slighted traditional geographical approaches while enthusiastically adopting insights and techniques from other fields, especially history and the social sciences. Let me explore and try to account for these tendencies.

Neglect of the traditional

Regional and descriptive geographies of the Caribbean are conspicuous by their absence. Only physical and historical geographers--William Morris Davis,[4] Charles Schuchert,[5] Carl Sauer[6]--have written about the entire area or its larger parts. Helmut Blume's West Indischen Inseln[7] is the only modern general study, and West and Augelli's Middle America[8] is the only up-to-date survey in English.

It is tempting to assume that so few studies exist because the daunting multiformity of Caribbean islands and mainlands confounds generalization. But if unmanageable diversity explains the absence of geographical overviews, it can hardly account for the paucity of local comprehensive surveys of clusters or even of individual islands and territories. Of a hundred islands and island groups that deserve full-scale inquiry, scarcely ten have been at all adequately treated: Cuba by Ortiz and Merrero,[9] Puerto Rico by Picó,[10] Guadeloupe by

---

[4]William Morris Davis, The Lesser Antilles (New York: American Geographical Society, 1926).

[5]Charles Schuchert, Historical Geology of the Antillean-Caribbean Region, or the Lands Bordering the Gulf of Mexico and the Caribbean Sea (New York: John Wiley, 1935).

[6]Carl Ortwin Sauer, The Early Spanish Main (Berkeley and Los Angeles: University of California Press, 1966).

[7]Helmut Blume, Die Westindischen Inseln (Braunschweig: Verlag Westermann, 1968); English translation in press (London: Longman's, 1973).

[8]Robert C. West and John P. Augelli, Middle America: Its Lands and Peoples (Englewood Cliffs, N.J.: Prentice-Hall, 1966).

[9]Fernando Ortiz, Cuban Counterpoint: Tobacco and Sugar (New York: Alfred A. Knopf, 1947); L. Marrero, Geografía de Cuba (Havana: Editorial Minerva, 1950).

[10]Rafael Picó, Nueva Geografía de Puerto Rico: Física, Económica, y Social (Río Piedras, P.R.: University of Puerto Rico, 1969). See also C. F. Jones and Rafael Picó, Symposium on the Geography of Puerto Rico (Río Piedras: University of Puerto Rico Press, 1955).

Lasserre,[11] Martinique by Refert,[12] Barbados by Starkey[13] (and that thirty years ago, primarily on the sugar economy), San Andrés and Providencia by Parsons,[14] St. Kitts and Nevis by Merrill,[15] Tobago by Niddrie.[16] No compendious studies exist of the major Commonwealth Caribbean countries, Jamaica, Trinidad, Guyana, Barbados beyond Starkey, though good secondary school geographies cover each of them and the group as a whole.[17]

Surveying the literature on the Caribbean, one wonders whether other scholars have preempted the geographers' traditional role. Island studies by anthropologists in particular abound, but a closer look at these works dispels the facade of territorial coverage. With a handful of exceptions--Julian Steward and others on Puerto Rico,[18] R. A. J. van Lier on Surinam,[19] M. G. Smith on

---

[11]Guy Lasserre, La Guadeloupe: Etude géographique, 2 vols. (Bordeaux: Union Française d'Impression, 1961).

[12]Eugène Revert, La Martinique: Etude géographique (Paris: Nouvelles Éditions Latines, 1949).

[13]Otis P. Starkey, The Economic Geography of Barbados: A Study of the Relationships between Environmental Variations and Economic Development (New York: Columbia University Press, 1939).

[14]James J. Parsons, San Andrés and Providencia: English-Speaking Islands in the Western Caribbean (Berkeley: University of California Publications in Geography, Vol. 12, No. 1, 1956).

[15]Gordon C. Merrill, The Historical Geography of St. Kitts and Nevis, the West Indies (Mexico City: Panamerican Institute of Geography and History, 1958).

[16]David L. Niddrie, Land Use and Population in Tobago: An Environmental Study (Bude, Cornwall: Geographical Publications, 1961).

[17]See, for example, F. Colin Evans, "First Geography" series on Jamaica, Trinidad, and the Eastern Caribbean (London: Cambridge University Press, 1972 and 1973); Wilma Williams-Bailey and P. H. Pemberton, Nelson's West Indian Geography (London: Nelson, 1972); R. M. Bent and Enid L. Bent-Golding, A Complete Geography of Jamaica (London and Glasgow: Collins, 1966); John Macpherson, Caribbean Lands: A Geography of the West Indies (London: Longmans, 1963).

[18]Julian H. Steward et al., The People of Puerto Rico: A Study in Social Anthropology (Urbana: University of Illinois Press, 1956).

[19]R. A. J. van Lier, Samenleving in een grensgebied: een sociaal-historische studie van de maatschappij in Suriname (The Hague: M. Nijhoff, 1949).

Grenada,[20] Leo Despres on Guyana[21]--these turn out to be studies of communities rather than territories: perfunctory introductions and conclusions link detailed analyses of particular localities to the history, geography, and institutional life of the territory as a whole. Only the historians have habitually dealt with islands as entities, incorporating a broad range of subject matter in their syntheses. But most of their studies, however holistic in content, limit themselves to particular periods of time.

The Caribbean has long attracted American and European geographers in search either of the exotic or of easily circumscribable areas suitable for sample studies and dissertations. But these attractive lures have yielded few exhaustive treatments of individual territories. Many scholars have set out to encompass a Caribbean island only to find its boundedness deceptive; even the smallest place may boast an abundance of significant data going back two to three centuries or more. To collect and synthesize these diverse materials proves a far more formidable task than anticipated, as I am all too well aware; in 1960 I embarked on general studies of Montserrat and Dominica; thirteen years later my cursory synopses of these small islands--40 and 300 square miles, 12 and 70 thousand people, respectively--now lies in partial drafts of several hundred typed pages, whose completion may require a lifetime.

Caribbean states themselves, despite their evident need for environmental information, have done little to encourage the collection and dissemination of geographical data. Only Guyana, thanks to the geographer Leslie Cummings, has mounted a national regional survey as an indispensible adjunct of proper resource utilization. Jamaica has a national atlas,[22] but it is based only on small-scale surveys, and until recently the Jamaican physical planning unit existed more on paper than in reality. On the other hand, the non-Hispanic Caribbean islands are fortunate in their basic map and air-photo coverage, legacies of British, French, and Dutch cadastral and ordinance surveys. Geo-

---

[20]M. G. Smith, Stratification in Grenada (Berkeley: University of California Press, 1965). See also M. G. Smith, Kinship and Community in Carriacou (New Haven: Yale University Press, 1962).

[21]Leo A. Despres, Cultural Pluralism and Nationalist Politics in Guyana (Chicago: Rand McNally, 1967).

[22]Jamaica Town Planning Department (Physical Planning Unit) and United Nations Development Programme, National Atlas of Jamaica, 2 vols. (Kingston, Jamaica: Government Printing Office, 1971). See also Colin G. Clarke, Jamaica in Maps (London: University of London Press, in press).

logical data there is also far superior to that in most of Latin America and indeed in much of North America.

A more severe deterrent to geographical synthesis than the plethora of raw data is the difficulty of knowing where, in substantive terms, to call a halt. In all parts of the world the borderlands between geography and history, the social, and the physical sciences are hard to manage; nowhere is it easy to draw a line between what belongs and what does not, between what is useful and what is extraneous. In the Caribbean these lines are particularly malleable. And the special Caribbean problems of content are a source both of confusion and of strength for geographical studies.

The Caribbean neglect of mainstream geography continues into the present. Many themes and motifs now current in America and Europe seldom appear in the work of Caribbean geographers. The quantitative wave has scarcely broken on these shores, as Parsons and Augelli note for Latin America too;[23] but in the Caribbean ecological concerns are also rare. Caribbean central business districts remain unstudied. Vernon Mulchansingh promises a spatial analysis of Trinidad based partly on Von Thünen,[24] and Ann Norton is investigating the hierarchy of central places in Jamaica,[25] but one searches in vain through the *Caribbean Geographer*[26] for other studies, completed or contemplated, that bespeak the jargon contemporary, or perhaps by now passé, in North America. Whether this is a shame or a blessing might best be decided by Caribbean anthropologists, recently afflicted with game-theoretical studies of family life in

---

[23] James J. Parsons, "The Contribution of Geography to Latin American Studies," in Charles Wagley (ed.), *Social Science Research in Latin America* (New York: Columbia University Press, 1964), p. 58; John P. Augelli, "Future Research in Latin America: Changing Professional Viewpoints and Issues," in Lentnek, Carmin, and Martinson (eds.), *Geographic Research on Latin America*, pp. 428-434.

[24] *The Caribbean Geographer* (University of the West Indies, Geography Department), Mona, Jamaica, No. 2 (1972), p. 28. See also Vernon Mulchansingh, "A Model Approach to the Understanding of the Transportation Network of Trinidad, West Indies," *Caribbean Quarterly*, Vol. 16, No. 3 (1970), pp. 23-53; Barry Floyd, "Planning for Rural Development in Jamaica: A Spatial Systems Analysis," *Caribbean Quarterly*, Vol. 18, No. 1 (1972).

[25] *Caribbean Geographer*, No. 2 (1972), p. 28.

[26] See Geography Department, University of the West Indies, Mona, Jamaica, "A Regional newsletter for geographers with teaching and research interests in the Caribbean," for a list of individuals and institutions concerned with the Caribbean, and their publications.

Martinique[27] and multiple regression formulae on attitudes toward modernization in the Bahamas and Trinidad.[28]

Multidisciplinary synthesis

Instead of sticking to their supposed last, Caribbean geographers have conspicuously transgressed disciplinary barriers, combining geography with insights from history, economics, and anthropology. Many social science studies of the Caribbean are not readily distinguishable by discipline. If there is a difference, it is that nongeographers tend to be more narrowly focused--anthropologists do community studies, economists commodity surveys, sociologists attitudinal studies, political scientists electoral studies--while geographers seek to synthesize these disparate specialties.

The Caribbean geographies noted above are as much historical as geographical, and are also heavily ethnographic, economic, or sociological in content. Several studies in progress display similar breadth: Colin G. Clarke's historical geography of Kingston, Jamaica,[29] is spatially focused yet treats the fabric of racial and ethnic distinctions temporally, as do his papers on East Indians in Trinidad.[30] Elizabeth Hope's study of Caribbean migration processes, owing much to diffusion studies of the Lund School, is economic and sociological in contents.[31] Barry Higman has gleaned data about the dynamics of slave

---

[27]Michael M. Horowitz, "A Decision Model of Conjugal Patterns in Martinique," Man, Vol. 2 (1967), pp. 445-453.

[28]William B. Rodgers, "The Wages of Change: An Anthropological Study of the Effects of Economic Development on Some Negro Communities in the Out Island Bahamas" (Ph.D. Dissertation, Stanford University, 1966); William B. Rodgers and John M. Long, "Male Models and Sexual Identification: A Case from the Out Island Bahamas," Human Organization, Vol. 21 (1968), pp. 326-331; F. T. Cloak, Jr., A Natural Order of Cultural Adoption and Loss in Trinidad (Chapel Hill: University of North Carolina Institute of Research in Social Science [Working Papers in Methodology No. 1], 1966).

[29]Colin G. Clarke, Kingston, Jamaica: Urban Development and Social Change, 1692-1962 (Berkeley and Los Angeles: University of California Press, 1974).

[30]Colin G. Clarke, "Caste among Hindus in a Town in Trinidad," in Barton M. Schwartz (ed.), Caste in Overseas Indian Communities (San Francisco: Chandler, 1967), pp. 165-199; "Residential Segregation and Intermarriage in San Fernando, Trinidad," Geographical Review, Vol. 61 (1971), pp. 198-218; "Pluralism and Stratification in San Fernando, Trinidad," Institute of British Geographers, Special Publication No. 5 (July, 1973), pp. 53-70.

[31]Elizabeth Hope, "Spatial Perception and the Environment in West Indian Migration," in preparation.

demography from early registration records as no other historian has done.[32] Like earlier geographers of the Berkeley school, these later Caribbean practitioners have found that geographical understanding must merge with other kinds. Geographers have, to be sure, a unique orientation, a skill in assessing space and place rare in other disciplines; I shall return to this in discussing opportunities to be tapped. But the essential point is that Caribbean geographers, past and present, not only have consciously borrowed insights and perspectives from scholarly colleagues in other fields, but have felt their own work to be in some ways inseparable from theirs.

This disciplinary fusion is a necessary consequence of the scale and of Caribbean affairs. The small size of discrete territories structures and sets in sharp relief the operations and interactions of social, economic, and political processes. Once scrutinized, each aspect of life is recognizably intertwined with every other sphere of activity. The interconnections are not only evident, they are locally prized as a network enabling interests to commingle. Because institutions are not compartmentalized, no single aspect of Caribbean affairs is forbiddingly recondite or abstruse--there is no apparatus of expertise meant to ward off the outsider. Unconcerned with professional mystique, many managers of Caribbean institutions are more or less inspired amateurs. Thus the geographer--or any generalist for that matter--comes to feel that nothing about Caribbean life is too arcane to learn; if he takes the time to absorb he can understand what is going on, not only in his own specialty but in Caribbean affairs generally.

This feeling that everything connects--so rare in Western societies--features local as well as foreign insights into Caribbean affairs. It is widely recognized, for example, that productivity in the sugar industry is a function not only of agronomic techniques, capital availability, and other economic and technical factors, but also reflects the domestic habits of migratory cane cutters, social distinctions between peasant farmers and estate owners, political sensitivity to absentee ownership as a colonial relic, and a rooted bias against sugar as the principle slave crop.

Personal history may best illustrate how some Caribbean geographers willy-nilly cross disciplinary boundaries. In Jamaica in the 1950's I joined a group of economists, anthropologists, sociologists, and historians at the Institute of Social and Economic Research of the University of the West Indies.

---

[32] Barry Higman, "Slave Population and Economy in Jamaica at the Time of Emancipation" (Ph. D. Dissertation, University of West Indies, 1971).

These scholars, West Indian, British, Canadian, South Asian, and Antipodean, gave me essential techniques and insights. An agricultural economist surveying small-scale farming in Jamaica made me aware of the social as well as the physical environment of Caribbean cultivators and the constraints of credit, insurance, marketing, and family needs that governed reactions to officially recommended innovations--the difference, in short, between the world as viewed by export firms, plantation managers, and government experts, and by rural folk caught up in a complex network of farm and nonfarm activities. An historian showed me how much Caribbean anti-authoritarianism stemmed from a colonial past with government by and for a European elite. A linguist taught me to appreciate how differences between formal English and folk patois affect inter-class, urban-rural, and inter-island relationships. An anthropologist made me aware of the dimensions of Caribbean race, class, and ethnicity, and of their interaction at local, national, and regional levels.

Each aspect of life threads into the seamless Caribbean social web. Thus curiosity about the high density of population in Barbados led me to compare demographic data. The apparent anomaly of low Barbadian growth rates reflected a high incidence of infant mortality paradoxical in the wider Caribbean health context. Medical records revealed local differences in parish administration, health services, and rural attitudes toward diet and hospitalization; community surveys showed that parental expectation of old-age support from _one_ of _many_ children was a mainstay of Barbadian domestic life. An inquiry into population distribution became in turn a demographic, a medical, a political, and a sociological study.[33] Similarly the local sense of relative well-being in overpopulated, eroded, basically impoverished Montserrat proved to be a consequence, in part, of the early demise of the white plantocracy, which left most black Montserratians without an elite to envy; in part of emigrant remittances; and in part of local self-government, giving Montserratians, by contrast with other small islanders, the feeling of being in charge of their own affairs, along with a built-in welfare system in the form of government employment.[34]

When I sought to learn why Dominica, largest and most luxuriant of the Lesser Antilles, was also the least productive and the most sparsely populated,

---

[33] David Lowenthal, "The Population of Barbados," _Social and Economic Studies_, Vol. 6 (1957), pp. 445-501.

[34] David Lowenthal and Lambros Comitas, "Emigration and Depopulation: Some Neglected Aspects of Population Geography," _Geographical Review_, Vol. 52 (1962), pp. 195-210.

I found the seeds of underdevelopment in colonial land speculation. A government anxious to build up British plantations against French and Carib settlers persuaded eighteenth century British entrepreneurs to buy huge, inaccessible, and sometimes non-existent tracts that could not be cleared or cultivated in time to gain title. Most tracts reverted to the crown and to squatter occupation, while the antecedent French and free-coloured population continued to outstrip the British super-structure. The social importance of land transcended its economic value, especially after emancipation, when property qualifications enabled coloured smallholders to outvote and gain legislative control from white planters. The so-called Mulatto Ascendency preferred Dominica's semi-wilderness status quo to development that might attract an influx of European planters, with political and racial consequences all too evident in the other islands. Meanwhile, a long succession of estate bankruptcies suffered by feckless or romantic entrepreneurs continues to make Dominica a graveyard for foreign investors. Local men of substance regard planting as a status-gratifying pastime, not as a paying enterprise.

If all this is history, anthropology, or psychology, it is no less clearly geography. In the situations I have delineated, individuals are directed and constrained both by actual circumstances and by their collective memories in landscapes that are palimpsests of the real and of imagined past and present. The temporal dimension is one that Caribbean geographers must especially reckon with. The West Indian tends to live in the past both consciously and unconsciously--he may be unaware of the specific historical forces that activate him, but he senses that he belongs, if not to history in a causal sense, to a stream of time, a temporal commingling of generations, genealogies, governments, and tenures. The social and physical langscape of a century past is recalled or reinvented as freshly as that of yesterday, and the present is reflected in the prism of that multiform past.

These, then, are some of the paths that Caribbean geographers have followed. Synthesists rather than specialists, they have taken the broadest possible views, assuming all knowledge as their province instead of concentrating on more limited tasks. This inclination, however, is nowadays decreasingly viable. Geographical work in the Caribbean promises to become more specialized and more programmatic. Cognate disciplines in the Caribbean, as elsewhere, are growing more arcane, institutionalized, and scientistic, and hence less accessible to geographers. The region's scholarly journals, once devoted to wide-ranging, substantive analyses illuminating Caribbean social and economic affairs, are now being replaced by theoretical, model-building abstractions, often couched in impenetrable social-science jargon, of little direct relevance to Caribbean realities.

Nor is local professional geography immune from academic institutionalization. The number of geographers working in the Caribbean tends to double each decade, and far more research is group-oriented and government-linked than formerly. The solitary Sauerian of the Berkeley School can still be found --Clarissa Kimber's recent garden studies in Puerto Rico and Martinique are representative examples[35]--but much research in progress is team work under such corporate auspices as McGill University's Savanna Research Institute and the Université de Montréal's Centre de Recherches Caraïbes.

The expansion of geography in Caribbean academic institutions further promotes pragmatism. Geography departments were inaugurated during the 1960's at the University of Puerto Rico, the University of Guyana, and the Jamaican campus of the University of the West Indies, the last now expanding to the Trinidad campus. National exigencies, university dependence on local governments, and increasingly local compositions of staffs and students all require research directed to problems of visible public concern. Caribbean geography departments, even more than those in American land-grant colleges, meet heavy demands to service local agencies--land authorities, agricultural offices, urban and rural planning units--with surveys, statistical data, and map compilations. The physical impacts and social implications of such major Caribbean industries as bauxite, oil, tourism, sugar, bananas and citrus also engage the attention of many local geographers. Under local auspices, pure research is supplanted by research as a vehicle for coping with issues of felt national urgency, as Augelli has similarly noted in connection with Latin America.[36]

Thus the professionalism of geography and related disciplines, coupled with the emerging nationalization of geography as an academic field, is leading research away from broad, humanistic syntheses toward applied, programmatic, service-oriented enterprises.

---

[35]Clarissa Kimber, "Dooryard Gardens of Martinique," Yearbook of the Association of Pacific Coast Geographers, Vol. 28 (1966), pp. 97-118; and "Spatial Patterning in the Dooryard Gardens of Puerto Rico," Geographical Review, Vol. 63 (1973), pp. 6-26.

[36]Augelli, "Future Research in Latin America," p. 432.

Potential Contributions of Geography to Caribbean Studies

One may hope that these avowedly practical exercises do not engage Caribbean geographers to the total exclusion of longer-range, fundamental contributions. Geographical insights seem to me especially applicable to Caribbean affairs in three broad realms of inquiry--matters of scale and location, of environment and habitat, and of connections with the wider world. Let me elaborate on each of these briefly.

Scale and location

The configuration of the Caribbean archipelago creates a congeries of circumstances whose just appreciation requires geographical insights and techniques of analysis. An underlying feature of all Caribbean milieus, otherwise so diverse, is their location within the self-contained land masses of this archipelago. Each island is a world in itself, as is so often said, and individuality is an inviolable virtue in the eyes of most West Indians. But territorial resemblances are likewise consequential; the general nature of the area is a product of the regularities as well as the local uniquenesses. Two circumstances make Caribbean assessments peculiarly difficult and also intrinsically geographical. One is the location of its various parts: the varying distances and topographic barriers that divide island from island, that link or separate territories of varying landform and climate, culture and ethnicity, language and imperial background. Thus Jamaica's proximity to the Hispanic islands and remoteness from other British Caribbean territories undermined the structure of the West Indies Federation. Thus Anguilla's easy access and environmental and socioeconomic similarities to French and Dutch St. Martin and the American Virgin Islands made her break from St. Kitts, only sixty miles away but altogether another kind of world, not only inherently logical but ultimately feasible. Thus Dominica's central position in the Lesser Antillean chain is weakened by the Gallic _cordon sanitaire_ of her immediate neighbors, Guadeloupe and Martinique; the French islands entail patterns of culture and economy that set Dominica off from the other British Windward and Leewards. Thus Barbados' easterly remove is an excuse for, rather than a feature of, Barbadian insularity; other islanders feel Barbados is alien by reasons of landscape and custom more than of location. Thus Trinidad's continentality is a function of landforms and social cosmopolitanism, along with proximity to Venezuela. These are the kinds of judgments that geographers ought to have particular talents for weighing.

Variations among Caribbean territories ranging from minuscule to quasi-continental affect the scale of life for both land and people. While clear physical and social demarcations make the Caribbean a superb setting for comparative analyses, entities of such varying size are not easy to compare. Haiti and the Dominican Republic share the same island and have roughly equivalent populations but wholly dissimilar histories and socioeconomic systems. Jamaica has more than twice the area and population of Trinidad, and Trinidad several times that of any other Commonwealth Caribbean island. There are only a few similar proximates: Guadeloupe and Martinique, the four British Windwards, St. Croix and St. Thomas in the American Virgin Islands. How then can comparisons be made? Is it valid to draw analogies between circumstances in an island of ten to twenty thousand people and those in one of a hundred thousand, let alone a million? Economic enterprise, political organization, social services, migratory patterns, and interaction networks necessarily differ for each order of magnitude. A dozen years ago I proposed a program comparing social structures in island-pairs of like size--Barbados and Guadeloupe, Antigua and Aruba, Grenada and St. Vincent, Anguilla and Carricou, St. Kitts and Tobago;[37] later I used Montserrat and Nevis to exemplify the method.[38] But this exercise now strikes me as barren, because the uncontrolled variables obviate the advantages of size equivalence. Differences of magnitude should not be overlooked in comparing Caribbean territories, but taken into account along with other variables. And this is a truly geographical task.

Environment and habitat

Most Caribbean peoples are today reevaluating local identities--regional, national, racial, and ethnic. After centuries of subordination, they have begun to be masters in their own house. But the sense of owning and belonging entails informed awareness of and devotion to local landscapes, local resources, and the sweep and specificities of the local past.[39] An instance of this awareness appears in a recent publicity handout celebrating the birth of the Co-operative Republic of Guyana:

> This grassroots revolution, . . . seems to engulf the psychic being of the Guyanese nation, . . . [with] a growing awareness of the beauty and power

---

[37] David Lowenthal, "The Range and Variation of Caribbean Societies," Annals of the New York Academy of Sciences, Vol. 83 (1960), pp. 786-795.

[38] Lowenthal and Comitas, op. cit.

[39] David Lowenthal, "Caribbean Views of Caribbean Lands," Canadian Geographer, Vol. 5, No. 2 (1961), pp. 1-9.

> of indigenous ideas, things, institutions and values . . . an enchantment that can only come from things Guyanese. . . . In the field of art there has emerged a new approach to things of the land . . . to recreate . . . in a world of aesthetic, the essentials of our roots.[40]

Yet to encourage such appreciation is a difficult enterprise. Caribbean experience has engendered the habit of rejecting local things and denying the significance of local history, the worth of local resources, the value of local landscapes. West Indians have traditionally despised what is West Indian. They have turned their backs on the Caribbean to embrace the supposed superiorities of Europe and America, and more recently of India and Africa.

Nevertheless, the Caribbean past is now emerging as a theme of recognized significance, if not yet one for self-congratulation. In the wake of decolonization and the black-power movement, scholars are reassessing the roles of masters and slaves, planters and laborers, Europeans and colonials, and especially of African roots. A leading part in this enterprise belongs to British Caribbean historians. Thus Edward Brathwaite has charted the everyday life of hitherto unchronicled Caribbean masses in both history and poetry.[41] More avowedly romantic, Cuban, Haitian, and French Antillean devotees of *négritude* variously commemorate peasantries and proletariats in poems, essays, and historical novels.[42] The new Caribbean history, like most new histories, is devoted to nationalist causes, but propaganda has not seriously perverted local scholarship. The abundance of source materials, the solid achievements of earlier antiquarians, and their own training and sophistication keeps Caribbean historians in touch with reality and discourages them from misconstruing fancy as fact.

The perceptive negations of V. S. Naipaul and others still persist--the sense that nothing of consequence has happened in the Caribbean, the feeling of

---

[40]"Thoughts on things within . . . At the roots," *Guynews*, Serial no. 1, Vol. 1 (November, 1971), p. 6.

[41]Edward Brathwaite, *The Development of Creole Society in Jamaica* (London: Oxford University Press, 1971); and *Rights of Passage, Masks, and Islands* (New York and London: Oxford University Press, 1967, 1968, 1969, respectively).

[42]G. R. Coulthard, *Race and Colour in Caribbean Literature* (London: Oxford University Press for the Institute of Race Relations, 1962), especially chapters I, II, IV; Alfred H. Berrian and Richard A. Long (eds.), *Negritude: Essays and Studies* (Hampton, Va.: Hampton Institute Press, 1967).

being cut off from the great world, the recognition of local impotence.[43] But West Indians are taking more and more interest in their Caribbean past as a key to understanding their present dilemmas.

No comparable claims can be made for the study of Caribbean lands and resources. Considered appreciation of place is more difficult to acquire than a regard for the past, for history can be recast in the mirror of present needs, whereas landscapes and townscapes are too ineluctably corporeal, too palpably intractable to be remade in new images. Notwithstanding a spate of new locally-oriented school texts, local environments are still remote from the minds and hearts of most West Indians. To the great majority, land and landscape are simply the scene of a necessitous scramble for livelihood; they are unaware of any passionate nexus of attachment to town or countryside. But geographers, too, have devoted little effort to tracing the history of relations between Caribbean man and land.[44]

Local rejection of local landscapes has several roots. European colonizers refused to view the Caribbean as a homeland and consequently made no effort to create livable environments. Other West Indians emulated this absenteeist mentality and learned that success meant quitting their own unloved landscapes. Meanwhile the most desirable environments and valuable resources were continually engrossed by outsiders. Caribbean landscapes are largely controlled, if not owned, by nonresident interests--international oil and bauxite companies, overseas plantation magnates, banks and real estate interests in New York and London, Montreal and Paris, and Amsterdam.

---

[43]See especially V. S. Naipaul, _The Middle Passage_ (London: Andre Deutsch, 1962), and _The Loss of El Dorado_ (London: Andre Deutsch, 1969).

[44]Outstanding among the scanty literature are David R. Harris, _Plants, Animals, and Man in the Outer Leeward Islands, West Indies: An Ecological Study of Antigua, Barbuda, and Anguilla_, University of California Publications in Geography, Vol. 18 (Berkeley and Los Angeles: University of California Press, 1965); and David Watts, _Man's Influence on the Vegetation of Barbados 1627 to 1800_ (Hull: University of Hull Occasional Papers in Geography, No. 4, 1966). See also Carl A. Carlozzi and Alice A. Carlozzi, _Conservation and Caribbean Regional Progress_ (Yellow Springs, Ohio: Antioch Press, 1968; Barry Floyd, "Conserving the Jamaican Environment: The Role of Geography," _Environmental Newsletter_ (Caribbean Conservation Association), Vol. 2, No. 2 (July, 1971), pp. 11-32; James J. Parsons, _The Green Turtle and Man_ (Gainesville: University of Florida Press, 1962); Gordon Merrill, "The Historical Record of Man as an Ecological Dominant in the Lesser Antilles," _Canadian Geographer_, No. 11 (1958), pp. 17-22; James M. Blaut _et al._, "A Study of Cultural Determinants of Soil Erosion and Conservation in the Blue Mountains of Jamaica: Progress Report," _Social and Economic Studies_, Vol. 8 (1959), pp. 403-420.

The prevalence of tourism throughout the Caribbean especially discourages links of affection between land and people. Tourist board paeans to sun and sea, to shining mountains and tropical verdure, are as remote from indigenous understanding as is tourist wealth from local impoverishment. To West Indians, landscape appreciation by outsiders takes on an exploitative flavor, connoting local exclusion from resource use. The usual stress between resident and visitor, farmer and vacationer, is exacerbated in the Caribbean by gross disparities in living standards and by distinctions of race and nationality between tourists and local inhabitants. To men locked into subsistence farming or urban underemployment, efforts to conserve beaches and protect bird life seem misplaced priorities analogous to the kodachrome syndrome that encourages strangers to imagine that because the Caribbean is lush and balmy life there must be easy. The picturesque view is the exotic image of visitors on vacation and of other birds of passage; the perspective is always European and American, never Caribbean. West Indians who are painfully aware of social realities argue that an interest in landscapes simply diverts attention from the human wretchedness of the people; to applaud the palm tree, as the Martiniquan Edouard Glissant puts it, is to neglect the miseries of those condemned to live beneath it.[45]

Contemporary Caribbean self-assertions--local sovereignty, the nationalization of resources, the eviction of suspect outsiders, the denial of land to aliens--are at odds with this historic rejection of local landscapes and resources, and may signal a heightened evaluation of that patrimony. A handful of West Indians--Wilson Harris[46] for Guyana, Derek Walcott[47] for St. Lucia and Trinidad, George Lamming[48] for Barbados and various mythical islands--have begun

---

[45] Edouard Glissant, La Lézarde (Paris: Editions du Seuil, 1958), quoted in Jack Corzani, Splendeur et misère: l'exotisme littéraire aux Antilles, Groupe Universitaire de Recherches Inter-Caraïbes, Études et Documents, No. 2 (June, 1969) (Pointe-à-Pitre, Guadeloupe: Centre d'Enseignement Supérieur Littéraire), p. 12.

[46] Wilson Harris's Guyanese novels include Palace of the Peacock (1960), The Far Journey of Oudin (1961), The Whole Armour (1962), The Secret Ladder (1963), Heartland (1964), The Eye of the Scarecrow (1965), Tumatumari (1968), and The Sleepers of Roraima: A Trilogy (1970), all London: Faber & Faber. See also his Tradition, the Writer, and Society: Critical Essays (London and Port of Spain: New Beacon Publications, 1967).

[47] Derek Walcott's poems are collected in In a Green Night: Poems 1948-1960 (1962), The Castaway and Other Poems (1965), and The Gulf and Other Poems (1969), all London: Jonathan Cape. His plays include Malcauchon, The Sea at Dauphin, The Dream on Monkey Mountain, and Ti-Jean and His Brothers.

[48] George Lamming's novels include In the Castle of My Skin (1953), The

to shape images of their homelands that are densely passionate, rich in metaphorical context, infused with the sights and sounds and smells of forest and foreshore, and embedded in folklore and folk speech. The plays, novels, and poems of these Caribbean writers offer the best available materials for a true Caribbean geography.

Geographers could contribute much to the emergence of this local awareness of place, to the forging of links between land and people. They might study sociocultural as well as economic uses of local resources, and show how landscapes are inhbited and traversed. Geographical perspectives should open out to encompass the social, cultural, and psychological milieus of Caribbean man. There are some beginnings in these directions. Donald Innis's[49] studies of small farms in Jamaica emphasize the multiple uses of land by household members, and their social and ecological as well as their commercial consequences. Clarissa Kimber[50] considers such desiderata for garden composition and layout as social space and esthetic preferences. Margaret Katzin[51] (an anthropologist of geographical bent) has followed the daily rounds of Jamaican higglers from home to customers to markets and back, in the context of buying and selling, to be sure, but also of kinship and comfort, distances and directions, isolation and gregariousness, tangible and intangible benefits and drawbacks of every kind. The sense of habitat and habitation, among other realms of environmental perception, calls for geographical insights. The use of local resources in the service of local interests; the marriage of land reform with rural planning and of urban growth and industrialization with town and city planning; an awareness

---

Emigrants (1954), Of Age and Innocence (1958), and Season of Adventure (1960), all London: Michael Joseph. See also his The Pleasures of Exile (London: Michael Joseph, 1960).

[49]Donald Q. Innis, "The Efficiency of Jamaican Peasant Land Use," Canadian Geographer, Vol. 5 (1961), pp. 19-23; "Problems of the Jamaican Peasant Economy," Spotlight (Kingston, Canada), Vol. 1 (1964), pp. 25-31; "Peasant Farming: Banana Intercropping in Jamaica," Proceedings of the Association of American Geographers, Vol. 5 (1973), pp. 111-117; "Coffee in Jamaica Peasant Intercropping," paper presented at the Association of Pacific Coast Geographers meeting, San Diego, June, 1973; "The Effects of Peasant Farming on Soil Fertility: A Jamaican Example," paper presented at the C. L. A. G. meeting, Calgary, June, 1973.

[50]Kimber, op. cit.

[51]Margaret Fisher Katzin, "The Jamaican Country Higgler," Social and Economic Studies, Vol. 8 (1959), pp. 421-440; "The Business of Higgling in Jamaica," Social and Economic Studies, Vol. 9 (1960), 297-331.

of man-environment interactions in the context of local ecology; landscape appreciation as an integral aspect of the Caribbean heritage--these are all tasks that commend themselves to Caribbean geographers in the years ahead.

Connections with the wider world

The obverse of Caribbean self-respect and indigenization is the continuance of no less vital external relationships. Caribbean territories are too small, too historically linked with old imperial centers, and too economically, culturally, and socially dependent truly to go it alone, as Haiti and Cuba in different fashions demonstrate. I have elsewhere[52] outlined these pervasive external connections under two summary headings: emigration, the more-or-less formalized patterns of departure from the Caribbean that West Indians have adopted and adapted to for centuries; and neocolonialism, the survival and recrudescence of Caribbean subordinations, notwithstanding political autonomy, and their social and psychological consequences.

Each of these broad rubrics requires further exploration and deeper analysis. For example, emigration needs to be assessed in the context of individual and group aspirations and achievements, both at home and abroad. External migration should be scrutinized as part of a broader movement from country to town and from island to island. Migrants, whether temporary or permanent, must be seen as strands in a comprehensive socioeconomic fabric that also includes those who have stayed at home. For a comparative study of island migration to Great Britain, the geographer Elizabeth Hope is currently investigating expectations in the home community and the extent to which they are borne out and transformed by the migration experience. Ongoing studies of Caribbean national groups in New York, Montreal, London, the Panama Canal Zone, and the American Virgin Islands--a way station to the American mainland--focus not merely on demographic statistics but also on individual life cycles and group dynamics. Geographers might begin to incorporate experiential dimensions, feelings, and attitudes in their work after the fashion of Oscar Lewis in Puerto Rico[53] and Robert Coles in Appalachia.[54]

They might also assess the significance of migration for environmental

---

[52] David Lowenthal, West Indian Societies (London and New York: Oxford University Press for the Institute of Race Relations, 1972), Chapter VI.

[53] Oscar Lewis, La Vida.

[54] Robert Coles, Migrants, Sharecroppers, Mountaineers: Volume III of Children of Crisis (Boston: Little, Brown, 1972).

experience and spatial learning, and the effects of emigration on local identity and national viability. To what extent, for example, does the continual trans-Caribbean and trans-Atlantic flux impel conservative stay-at-home islanders toward open and innovative societies? To what extent, *per contra*, does that movement constrain local energies and dissipate loyal loyalties? What are the realities, and how far do they depart from local ideals and expectations, of migrant contributions to Caribbean styles of life and thought?

No less compelling are questions concerning neocolonialism. Despite the ferment of radical geography, with its denunciations of imperialism and its commitment to the powerless and the underprivileged, few geographers appear to be interested in the effects of the imperial order in such areas as the Caribbean.[55] The tyranny of capitalism and the subordination of Third World peoples are assumed as wrongs to be righted; the tasks of assessing forces and mechanisms, of understanding causes and consequences, and of surveying the social and psychological impacts of economic control, are left to other social scientists. With rare exceptions, social advocacy among geographers produces polemics rather than fruitful research.

Thus the Caribbean impact of multinational corporations, whose pervasive power Raymond Vernon documents,[56] has been ably studied by Caribbean economists--George Beckford for commercial agriculture,[57] Norman Girvan for bauxite and for foreign capital generally,[58] Clive Thomas for banking,[59] and

---

[55] For exceptions revealing, see William Bunge, "Exploration in Guadeloupe: Region of the Future," *Antipode*, Vol. 5, No. 2 (May, 1973), pp. 67-83; and Curtis C. Roseman *et al.*, *Social Problems in a Small Jamaican Town* (Champaign: Univ. of Illinois at Urbana, Geography Graduate Student Association (Occasional Publications of the Dept. of Geography, Paper No. 6), 1973. See also L. Alan Eyre, "The Shanty Towns of Montego Bay, Jamaica," *Geographical Review*, Vol. 62 (1972), pp. 394-413.

[56] Raymond Vernon, *Sovereignty at Bay: The Multinational Spread of U.S. Enterprises* (New York: Basic Books, 1971).

[57] George L. Beckford, *Persistent Poverty: Underdevelopment in Plantation Economies of the Third World* (New York and London: Oxford University Press, 1972).

[58] Norman Girvan, *The Caribbean Bauxite Industry* (Jamaica: Institute of Social and Economic Research, University of the West Indies, 1967), and *Foreign Capital and Economic Underdevelopment in Jamaica* (Jamaica: Institute of Social and Economic Research, University of the West Indies, 1971).

[59] Clive Y. Thomas, *The Structure, Performance and Prospects of Central Banking in the Caribbean* (Jamaica: Institute of Social and Economic Research, University of the West Indies, 1972).

others. In this realm geographers have scarcely gone beyond the descriptive surveys of Starkey, Kingsbury, and Fentem.[60] External impacts on Caribbean culture and mores, institutions and values, are generally left to sociologists and anthropologists, though many of these topics are supremely geographical. Who, other than Ray Crist, has looked at inter- and intra-island distinctions of paysage and genre de vie?[61] Who has examined the diffusion of material goods, techniques, ideals, and stereotypes--European, American, Indian, African--from class to class, community to community, or ethnic group to ethnic group within the Caribbean? When will a geographer look at the Trinidad steel band, now a pan-Caribbean symbol, to see how national promotion and international adaptation affect the distance-decay of local innovations?

This classification of external influences is itself ad hoc, artificial, and inadequate. New research exploring how Caribbean global contacts penetrate and interact with local activities and allegiances will yield other syntheses. An informed understanding of the Caribbean, both as a general realm and as a constellation of individuals in states and communities, will have to weigh such conflicting impulses as contact and isolation, cultural borrowing and local invention, going abroad and coming home (whatever "home" may be), and the local consciousness of these dualities.

Perhaps the single most crucial element in the Caribbean scene today is the tension between territorial and social identity. Centuries of colonial history persuaded West Indians that they were primarily British, French, Spanish,

---

[60] Arlin D. Fentem, Commercial Geography of Dominica (1960), Commercial Geography of Antigua (1961), Commercial Geography of St. Vincent (1961); Robert C. Kingsbury, Commercial Geography of the British Virgin Islands (1960), Commercial Geography of the Grenadines (1960), Commercial Geography of Trinidad and Tobago (1960), Commercial Geography of Grenada (1960); Otis P. Starkey, Commercial Geography of Montserrat (1960), Commercial Geography of Barbados (1961), Commercial Geography of St. Kitts-Nevis (1961), Commercial Geography of St. Lucia (1961), Commercial Geography of the Eastern British Caribbean (1961), all Bloomington: Indiana University. See also F. L. Oxtoby, "The Role of Political Factors in the Virgin Islands Watch Industry," Geographical Review, Vol. 60 (1970), pp. 463-474.

[61] Raymond E. Crist, "Static and Emerging Cultural Landscapes in St. Kitts and Nevis, British West Indies," Economic Geography, Vol. 25 (1949), pp. 133-145; "Cultural Dichotomy in the Island of Hispaniola," Economic Geography, Vol. 28 (1952), pp. 105-121; "Changing Cultural Landscapes in Antigua, B.W.I.," American Journal of Economics and Sociology, Vol. 13 (1954), pp. 225-232. See also John P. Augelli, "Patterns and Problems of Land Tenure in the Lesser Antilles, Antigua, B.W.I.," Economic Geography, Vol. 29 (1953), pp. 362-367.

Dutch--in short, European. The common denominator of Caribbean experience was imperially-induced schizophrenia: all eyes on the distant metropolis, blindness and unconcern about nearby Caribbean connections and similarities. And when Caribbean likenesses were recognized they engendered not solidarity but hostility--Puerto Rican contempt for Cubans and their language, Martiniquan scorn for the "savages" of Guadeloupe, Jamaican concern about subversion from Cuba and Haiti, inter-Guianese relations confined to geological exploration and the return of escaped convicts and lepers.

Differences and rivalries remain impediments, but colonial devolution is now dissolving political and psychological barriers to Caribbean interaction. More serious are the barriers of tribalistic feeling. The assertion of national and territorial identities, the consciousness of being Guyanese or Jamaican or Curacaoan, both activate and conflict with ethnic, racial, and class antagonisms, setting group against group within each territory. These rivalries are not new, they were inherent in the old colonial order; but they are surfacing today with greater magnitude than ever. As a result, the composite social amalgam suffers defections by all its constituents--blacks, whites, East Indians, Chinese, the middle classes--historically deprived majorities and threatened minorities alike. The emerging character of the Caribbean as a region and the quality of life in that region are likely to hinge on the struggle between two sets of contrary impulses, the one toward territorial integrity and national identity, the other toward a tribalism of race, class, and ethnic group.[62]

Identities of place involve choices among loyalties at different scales--the archipelago, the island, the parish, the village. Yet most of these are small enough to conflict with other allegiances based on racial or ethnic tribalism, where each such loyalty involves a linkage with a remote Old World homeland, sundering the insular fabric itself. The choice between place and race cannot be foretold. But it must be a vital concern of Caribbean geography.

---

[62]The geographical literature on Caribbean race and ethnicity--not to mention that from the other social sciences--is too vast to summarize here. I have surveyed portions of it in my West Indian Societies and in "Race and Color in the West Indies," Daedalus, Vol. 96 (1967), pp. 580-626. National, territorial, and regional identity are treated in the above and also in my "Levels of West Indian Government," Social and Economic Studies, Vol. 11 (1962), pp. 363-391, "The West Indies Chooses a Capital," Geographical Review, Vol. 48 (1958), pp. 336-364, and The West Indies Federation: Perspectives on a New Nation (New York: Columbia University Press for the American Geographical Society and Carleton University, 1961).

CHAPTER 4

THE PACIFIC REALM

H. C. Brookfield
McGill University

The central question tackled in this paper is the effectiveness of geographers' work in contributing to regional study. This question has two parts: the first concerns our effectiveness in contributing to understanding, that is to say essentially our contribution to "science" as such; the second concerns our effectiveness in advancing knowledge that is of value in the present and early future in a more public area than the scientific, that is to say essentially the relevance of our contribution. Before I approach these questions, however, I first define the region I am discussing, then analyze briefly the history of geographical and other inquiry within that region. This will set the scene for the evaluation with which the paper is concluded.

What Pacific Realm?

We have two "marine" realms in this symposium, presenting us with conceptual problems of a rather different order from the discussion of the role of geography on contiguous land. The accusation levelled at geographers that we are in fact "terrographers" is particularly relevant here, for the unity of an oceanic space is marine and the seas must form part of our discussion. Yet we do not speak of an "Atlantic realm" or even an "Indian Ocean realm": the Caribbean and the Pacific have a more positive perceived unity than either of these. Both have a large regional literature, including several active journals, research organizations and popular news-magazines and newspapers. As a first task we must explore this regional unity, and seek to understand its bounds. These are by no means self-evident, for there are several orders of "Pacific realm." In a larger sense, that of the Pacific Science Association for example, all countries bordering on the ocean are included together with countries bordering on adjacent seas in southeast Asia. In another sense there are only the island regions and oceans away from the surrounding mainlands and the larger archipelagoes of southeast Asia--but although this definition is often used, New Guinea is always included. It will aid clarification to spend a little time considering this question of definition as a problem in human geography.

At least until the present century, the Pacific realm existed solely as a construct of European colonial expansion. Though Polynesian voyagers extended their activities through several centuries over a huge area from New Zealand to Hawaii and to Easter Island, they never maintained integrated networks of regular movement outside a much more restricted belt lying within the tropics and south of the equator, within which no island is separated by more than 500 kilometers from some neighbour. Accidental voyages apart, they rarely voyaged further than from one island group to the next: contact was carried through the whole region, but not by the same mariners, so that the far east knew nothing of the far west. As European navigators explored the ocean, reducing the hypothetical Terra Australis Incognita to the present Australia and New Zealand, together with a festoon of islands spread out in a great arc to the north and east, they discovered four or five main groups of indigenous people who for centuries or millennia had known minimal contact with the surrounding continents. Though Asian voyagers certainly reached the Marianas, New Guinea and northern Australia, there is no evidence that they ever went further and returned to tell the tale. Between 1788 and 1906 the whole of this region, all but a few islands occupied from Japan, fell under European colonial rule without sustained or concentrated opposition such as faced the European advance into Asia or the ancient empires of tropical America. Large and important parts--Australia and New Zealand, Hawaii, New Caledonia, and also tiny Pitcairn, became colonies of European settlement: already by the mid-nineteenth century most of southeastern Australia was settled, and the rape of Polynesian New Zealand and Hawaii were well advanced; by the end of the century the process of occupation by settlement had reached its effective limits, and commercial colonization of the inner island regions was proceeding apace. Subsequently, in the present century, Japan succeeded in taking control of quite large areas from European occupiers, mostly for only a few short years in the early 1940's, but elsewhere for thirty years; Japanese commercial penetration has subsequently returned in force, and spread to areas never controlled by Japan as a colonial power. At the same time, Australia and New Zealand also became colonial powers in the region, and remain so to this day. The steady replacement of the European powers by the riparian nations in political, cultural, and commercial intervention in this ocean has been continuous since the beginning of this century.

From a human point of view, then, there are two contrasted ways of viewing and discussing the contemporary Pacific realm. One is the "cultural" approach, distinguishing regions within the Pacific basin according to the dominant culture complex--which opens some difficult questions in areas of very mixed population, such as Hawaii, New Caledonia, Fiji, and northern Australia.[1] The major groups in such an approach would be the "traditional" Melanesia, Polynesia, and Micronesia, together with mainly-European Australia and New Zealand[2] and a fringe of Asian culture in the north and west. The alternative approach is to examine the region in terms of stage and nature of development.

We could approach the problem with the tools of modernization theory, as evolved by Deutsch, Eisenstadt, and others, and rendered into spatial terms by Gould and Soja. If we could commit the cartographic sin of linking up pieces of land by iso-lines across water, we would obtain results that would differ very much according to the scale of our examination. Over the whole ocean basin we would have poles of highest modernization in North America, Japan, Australia, and New Zealand, and the slope of the modernization surface would decline to a trough running through the tropical Pacific to a deep sink in west New Guinea. Enlarging the scale, we would find within this trough a pattern of small growth centers around points like Suva, Papeete, Port Moresby, and especially Nouméa, with a decline away to periphery around them.

This purely mental simulation points up the fact that we have around the Pacific basin some of the world's wealthiest countries, and also all three of the world's most powerful. Accustomed as we are to thinking in terms of a world centered around the North Atlantic basin--or on points west or east of this basin, depending on viewpoint--it is arresting to recall that a 70° _calotte_ of the earth's surface, centered at the middle of New Guinea, will include all of the territory of Australia, New Zealand, and Japan, and parts of China, the Soviet Union and USA (Hawaii). Perhaps the most fundamental problem of Pacific geography then is the largest: in the center of all this wealth and power lies a scattered, sparsely distributed, relatively unpopulous region that is almost in its entirety thoroughly underdeveloped.

---

[1]What would we say of the dominant "culture" on present-day Guam on this basis?

[2]For a discussion of the origins of these terms, and their areal extension, see H. C. Brookfield with D. Hart, _Melanesia: A Geographical Interpretation of an Island World_ (London: Methuen, 1971), pp. xliii-xlv. In their present form, they date from classification by French writers in the 1830's.

There is virtue in this sort of old-fashioned geopolitical view, because it helps us to distinguish the central problems of a region. If we compare this Pacific situation with that of the Caribbean, we see that it is not parallel. There is the might of the USA towering over the region to the north[3] and there is Europe, but on the other side is the turbulent and largely underdeveloped continent of Latin America. To the Pacific islanders, Latin America is so remote as scarcely to exist: the only proximate comparable area is southeast Asia, and only for a short period in Sukarno's day has this yet been seen in the Pacific as a presence and a force.

The view of the Pacific thus varies substantially according to the spot from which we take it. American geostrategists who, like the British before them, like to see maps coloured in definite hues, see almost the whole ocean including Australia[4] in a satisfying shade, with only a grey zone over on the western border. This is a legitimate view, for almost the whole ocean does lie open to American commercial enterprise, tourism, and strategic design. A Japanese view is more likely to see a wedge widening southward from their islands, incorporating much of southeast Asia, the Pacific islands, and at least northern Australia, in which Japanese enterprise is currently growing more and more dominant--a sort of greater Greater East Asia Co-prosperity Sphere revamped in modern terms. An Australian view sees danger to the northwest, in Asia, a valuable but aggressive trading partner in Japan, and a kind of broad arc running east from Malaysia far across the Pacific in which their trading interests are variably dominant; the Pacific sector of this is brightening, becoming collectively a major trading partner in recent years. Given the difference of diminishing scale, all these views have much in common. But the Pacific Islanders' view is different. They see an island home of ancient culture more and more deeply invaded by foreign enterprise and cultural modernization. The trough and sink of the modernization surface are their home, and while they wish to improve their lot, they do not wish this to be at the price of a total loss of identity. They hope to achieve rapid growth in order to attain strength; they are not willing to pay the price of absorption.

---

[3]And the geographers of the USA do little to diminish this sense of overpowering: in their statement in The Caribbean Geographer, Vol. 2 (1972), the Association of American Geographers open by reminding Caribbean geographers that "they may participate in the activities of the A. A. G. by becoming members of the Southeastern Division of the organization." Canadians get like treatment.

[4]At least until December 1972.

The Island Realm

I therefore concentrate on the island realm, though I shall stress its interconnections with the "centers" grouped around its periphery. I shall examine briefly the distinctive structure of scientific inquiry in the island realm, and the role of geographers. I shall then turn to the question of how well we have served the islanders themselves in all this work, and hope to indicate some paths for the future.

In the sense that the writings of Cook, Banks, and the Forsters are part of the evolution of geography from mere description into a scientific discipline, the Pacific islands are also an integral part of that early history. Through the voyage of Darwin in the Beagle, and the excellent writings of d'Urville, Wilkes, and others among the last true explorers of the Pacific, this link is carried forward into the nineteenth century, though weakly, for the German pioneers of a more modern geography sought their material closer to home. And thereafter it was a long time before the Pacific contributed much more to geography.

Through the second half of the nineteenth century the evolution of the modern disciplines, following the impact of Darwin, took place mainly within the confines of Europe and North America. Scientific writing beyond these realms was extension of knowledge, rather than its analysis. Thus one of the earliest--perhaps the earliest--of Pacific "geographies," C. E. Meinicke's Die Inseln des Stillen Oceans: eine geographische monographie (1875-76, republished Amsterdam, 1969, 2 vols.), is fundamentally an anthology of discovery and exploration. But out of this sort of exploratory research came the evolution of a discipline which continues to play a dominant role in Pacific science. Pioneers such as Haddon and Seligman laid much of the foundation of modern anthropology, which then took a decisive step forward through the researches of Bronislaw Malinowski in Papua--very largely making a virtue out of his enforced stay in the field in the early part of World War I. He has been followed by a growing stream--Mead, Fortune, Firth, Hogbin, Emory, Piddington, and many others--so that "anthropological theory was evolved very largely in Oceania, where the relative isolation of small communities provided something like 'typical' primitive social groups."[5] Anthropology was the first social science to move outside a primarily European context of work: its earliness led to an isolation that has perhaps been

---

[5] A. I. Richards, "The Village Census in the Study of Culture Contact," Africa, Vol. 8 (1935), p. 20.

unfortunate for the discipline in several ways that will become relevant to our argument below.

Others who grew interested in the Pacific for its own sake, however, included certain natural scientists for whom the Pacific offered an immense laboratory for the study of such phenomena as coral growth and atoll formation, vulcanism, ocean currents and the dispersal of plants and animal life. The problem of the origin of the Pacific and its mountain and island systems presented itself to earth scientists, and serious and organized research in these areas began at about the same time as with the anthropologists, in the years around the turn of the present century. The cross-disciplinary nature of many problems, and the logistic advantages of combining research in several specialisms in a single costly expedition, encouraged collaboration, and several attempts were made between 1898 and 1916 to organize joint discussion and research on Pacific problems. The latter date is of some significance, for it marks a symposium on "Pacific exploration" organized by W. M. Davis, which led to the appointment of a Committee on Pacific Exploration by the U.S. National Academy of Sciences. In 1919 a meeting of the Pacific Division of the American Association for the Advancement of Science was held at Honolulu, and the work of Davis' committee, brought before this conference, led to an international scientific meeting in Honolulu in the next year. This became, by the *post-facto* setting up of a continuing organization, the first Pacific Science Congress.[6]

The pattern slowly widened through subsequent conferences, in Australia, Japan, Java, British Columbia, and California, before World War II. The emphasis remained on the natural and biological sciences, and on anthropology

---

[6]The principal topics discussed at this meeting were as follows:

- Race relations of the Pacific
- Program for anthropological research in Polynesia
- Origin of Hawaiian fauna and flora
- Fisheries of the Pacific
- Biological institutions in the Pacific region
- Program for work in Pacific volcanology (*sic.*)
- Seismology of the Pacific
- Mapping the Pacific
- Pacific ocean currents in relation to organisms
- Status of areal geological mapping in the Pacific region
- Correlations of post-cretaceous formations in the Pacific region
- Symposium on means and methods of cooperation
- Symposium on training of scientists for Pacific work

"Proceedings of the First Pan-Pacific Scientific Conference," *Bernice P. Bishop Museum Special Publication 7* (Honolulu, 1921).

together with certain related social sciences. The reports on these meetings are concerned with mainly technical matters, but with growing emphasis on the use of resources, and on human problems such as the depopulation which was then widely regarded as the overriding question concerning the "native" population. As Malinowski put it in 1921, and his view was echoed by many in the subsequent two decades:

> Ethnology is in the sadly ludicrous, not to say tragic, position, that at the very moment when it begins to put its workshop in order, to forge its proper tools, to start ready for work on its appointed task, the material of its study melts away with hopeless rapidity. Just now, when the methods and aims of scientific field ethnology have taken shape, when men fully trained for the work have begun to travel into savage countries and study their inhabitants--these die away under our very eyes.[7]

From this concern came a growing emphasis on public health, and on the conservation of people as well as of resources. And so gradually, but inevitably, came a shift toward questions of development, and of man in his environment.

World War II brought about a major change. Many geographers and other scientists now visited the Pacific, mostly at rather early stages in their careers, and at the same time there was a sudden demand for scientific information that was simply unavailable. The "Terrain Studies" hastily rustled up ahead of the campaigns, or the more academically precise four-volume "Geographical Handbook" of the British Naval Intelligence Division--prepared very largely by an historian[8]--provided basic information never previously collected and published. The Naval Intelligence Division "Handbook" remained a basic source for more than a generation, and is still valuable today. The aftermath of this war brought the atomic bomb, the Bikini experiments, and scientific exploration into what lay beneath Bikini--and other atolls. The Atoll Research Program, which still continues, began its major exploration of the geology, marine and subaerial biology, and morphology of atolls, yielding data of major importance in the evolution of ecosystem theory in the later 1950's and 1960's. Evidence from islands, of importance to biogeography since Darwin's sojourn in the Galápagos and even before, now came to be fed into a growing general theory of evolution and migration of species.[9] These developments followed directly from pre-war research,

---

[7] B. Malinowski, Argonauts of the Western Pacific (London, 1922), p. xv.

[8] Naval Intelligence Division, Geographical Handbook Series: the Pacific Islands (London: Admirality, 1943-45), 4 vols.

[9] See, for example, R. H. MacArthur and E. O. Wilson, The Theory of Island Biogeography, Monographs in Population Biology 1 (Princeton: Princeton University Press, 1967).

expanded by the newer ideas and greater resources of the post-war period. So also did the expansion of anthropological research that took place swiftly after 1946. But in other respects much happened that was wholly new.

The structure of scientific inquiry in the Pacific was changed greatly by a number of events that took place in the 1950's. One was the expansion of the small University of Hawaii, and the establishment on its campus of the East-West Center. Another was the foundation of the Australian National University, effectively from 1951, incorporating from the outset a Research School of Pacific Studies. A third was the formation at the end of the 1940's of the South Pacific Commission, an inter-governmental body providing scientific and advisory services, and carrying on a limited measure of basic research. And even more fundamental was the rising tide of anti-colonial revolution in Asia and then in Africa, causing more and more field researchers to turn to a Pacific Ocean in which the stable forms of colonial government still provided a kindly and undemanding environment for research--notwithstanding the difficulty experienced by a minority of middle-of-the-road liberals in gaining entry because of their "neo-communist" affiliations. There was something approaching an exponential growth in the volume of research in the Pacific right through the two decades from 1950 to 1970. It included geographers, and after a successful preliminary dêbut at Bangkok in 1957, these participated substantially and as a Section in their own right, for the first time in the 1961 Pacific Science Congress at Honolulu.

A Stocktaking in 1961

It is coming more and more clearly to be realized that the two week meeting of the Pacific Science Association in Honolulu in August 1961 represents not only the major stocktaking of the immediate post-war period, but also the origin of a number of developments that have become of growing importance in the subsequent decade. It can be said of this conference, as of few others of its size and scope, that it was a major scientific event, at once a bringing together and a point of departure.

Unusually also, some of the most important events were the Congress symposia, and most particularly a UNESCO-funded symposium on "Man's Place in the Island Ecosystem"[10] which brought together for the first time a

---

[10] F. R. Fosberg (ed.), Man's Place in the Island Ecosystem: A Symposium (Honolulu: Bishop Museum Press, 1963).

new generation of social scientists of the Pacific within a framework conceived by natural scientists. As speakers or as discussants, ten geographers and five anthropologists participated in this symposium, which is acknowledged as an important stage in the evolution of a modern and dynamic approach to cultural ecology.

The development thus signalled had been under way for some years, and the point that is critical to my argument is that its evolution in the Pacific owes as much to geographers as it does to anthropologists. The real beginning of this geographical development was in New Zealand, where K. B. Cumberland's department at Auckland initiated work very different in scope and especially scale from the generality of geographical inquiry in the late 1940's and early 1950's. The external inspiration came from L. Dudley Stamp and his pioneer Land Utilization Survey of Britain in the 1930's, as is made very clear in one of the earliest general papers.[11] Of necessity, the study of indigenous land use on Pacific islands involved work at a very detailed scale. In accordance with the ideas of the time, the scope of research remained territorial, but territories such as the Cook Islands, for example, are very small.[12] In the middle fifties, the Auckland department obtained a grant for a large scale attack on the problems of land use and development in Western Samoa[13] which included some very close inquiry at the village level. Other New Zealand geographers of the same formation have gone on to do similar work in Fiji,[14] the Solomon Islands and elsewhere in the Pacific.

The New Zealanders were not alone: in any case the diaspora of New Zealand geographers--of whom there are certainly more at work overseas than in New Zealand--ensured a dispersal of their own enterprise. There were some American contributions to work of this order, including one rather remarkable

---

[11] K. B. Cumberland, "Geography and Land-Use Survey in the Southwest Pacific," New Zealand Geographer, Vol. 5 (1951), pp. 1-18.

[12] W. B. Johnston, "The Citrus Industry of the Cook Islands," New Zealand Geographer, Vol. 7 (1951), pp. 121-138; "Land, People and Progress in the Cook Islands," Economic Geography, Vol. 29 (1953), pp. 107-124.

[13] J. W. Fox and K. B. Cumberland (eds.), Western Samoa: Land, Life and Agriculture in Tropical Polynesia (Christchurch: Whitcombe and Tombs, 1962).

[14] E.g., R. G. Ward, Land Use and Population in Fiji: A Geographical Study (London: H.M.S.O., 1965); R. F. Watters, Koro: Economic Development and Social Change in Fiji (Oxford: Clarendon Press, 1969).

one by R. E. Murphy,[15] which is much out of character with his later work. Also, about the same time, appeared Herold Wiens' largely, but not entirely, natural scientific collation of knowledge on atolls.[16]

More especially interesting among American contributors, however, was the new wave of ecologically-oriented anthropologists, perhaps beginning with Leonard Mason's studies in the Marshall Islands[17] and with the theorizing of Ward Goodenough on the interrelation of kin structures and the transmission of land rights to land resources.[18] But it was H. C. Conklin and his group at Columbia who truly opened the new field--and Conklin, it should be recorded, was once a student of Carl Sauer at Berkeley. Vayda and Rappaport, now two of the leading workers in this field of research and teaching, made their debut in these years, and especially at the Honolulu Congress. Rather less happy was Marshall Sahlins' first, highly deterministic, essay into the ecological field in the Pacific; his later analysis of a single island in Fiji was of a wholly different order,[19] and has become a classic work.

There were also the Australians--at this stage almost entirely a small group of geographers, anthropologists, historians, and later prehistorians in the Australian National University. The initial approach to Pacific research by the geographers drew very much on the New Zealand model, though there was also a direct involvement in public policy matters that always distinguished the work of this School in Canberra, and at times became very productively

---

[15] R. E. Murphy, "The Economic Geography of a Micronesian Atoll," Annals of the Association of American Geographers, Vol. 40 (1950), pp. 58-63.

[16] H. Wiens, Atoll Environment and Ecology (New Haven: Yale University Press, 1962).

[17] L. Mason, "The Bikinians: A Transplanted Population," Human Organization, Vol. 9 (1950), pp. 5-15; "Ecologic Change and Culture Pattern in the Resettlement of Bikini Marshallese," in V. Ray (ed.), Cultural Stability and Cultural Change (Seattle: American Ethnological Society, 1957), pp. 1-6.

[18] W. H. Goodenough, "A Problem in Malayo-Polynesian Social Organization," American Anthropologist, Vol. 57 (1955), pp. 71-83.

[19] A. P. Vayda, "Polynesian Cultural Distributions in New Perspective," American Anthropologist, Vol. 61 (1959), pp. 817-828; A. P. Vayda and R. A. Rappaport, "Island Cultures," in F. R. Fosberg, op. cit., pp. 133-142; R. A. Rappaport, "Aspects of Man's Influence upon Island Ecosystems: Alteration and Control," ibid., pp. 155-174; M. Sahlins, Social Stratification in Polynesia (Seattle: University of Washington Press, 1958); Moala: Culture and Nature on a Fijian Island (Ann Arbor: University of Michigan Press, 1962).

explicit.[20] Departure from the model arose directly from the close-meshed interdisciplinary structure of the Research School of Pacific Studies, and can perhaps best be identified as a shift of emphasis from land toward man in subject matter, and away from differentiation toward process in *explicandum*. The shift was by no means immediate, and land-use mapping remained for some time a fundamental research tool,[21] but collaboration with other disciplines led to changes in the research question. The ground-plan of the social structure of society became a central, rather than merely peripheral interest; welfare of man rather than of land became the main concern; the whole development of society came to override the development of resources; population geography took on a larger demographic element.[22]

Also of major importance in this new stream of work was the contribution of the ethnobotanists and prehistorians. In the modern sense, both were almost wholly new disciplines in the Pacific region, and so also was the mélange of work using documentary, oral, and indirect sources that was becoming "Pacific history." In a region where "prehistory" runs forward in some areas into the lifetime of living men, these several and seemingly separate disciplines are closely intertwined. Thus in his numerous studies of the agronomy of traditional Pacific agriculture, Barrau[23] also created the documents for comparative study

---

[20] O. H. K. Spate, "The Fijian People: Economic Problems and Prospects," *Fiji Legislative Council Paper 13 of 1959* (Suva: Government Printer, 1959). The earlier pattern, closely according with contemporary work by the New Zealanders, is best exhibited in J. N. Mercer and P. Scott, "Changing Village Agriculture in Western Samoa," *Geographical Journal*, Vol. 124 (1958), pp. 347-360.

[21] P. Brown and H. C. Brookfield, "Chimbu Land and Society," *Oceania*, Vol. 30 (1959), pp. 1-75 had a land-use map as its basic document, but in the second major field report by these authors--H. C. Brookfield and P. Brown, *Struggle for Land: Agriculture and Group Territories among the Chimbu of the New Guinea Highlands* (Melbourne: Oxford University Press, 1963)--land-use mapping gave way to the mapping of group territory both in detail and in extenso.

[22] For example, among the unpublished Ph. D. theses at A. N. U., those of D. R. Howlett, *A Decade of Change in the Goroka Valley, New Guinea: Land Use and Development in the 1950s* (1962); P. N. D. Pirie, *The Geography of Population in Western Samoa* (1963); D. A. M. Lea, *Abelam Land and Sustenance: Swidden Horticulture in an Area of High Population Density, Maprik, New Guinea* (1964); A. M. Maude, *Population, Land and Livelihood in Tonga* (1965).

[23] J. Barrau, "Economie vivrière traditionelle et progrès agricole en Melanésie," *Bulletin du Pacifique Sud*, Vol. 4 (1954), pp. 1-13; "Le Milieu et l'agriculture traditionelle en Melanésie," *Annales de Géographie*, Vol. 65 (1956),

in the present by geographers and anthropologists, and with the evidence of the past by prehistorians. Perhaps more than any other single element, this work has been central to the fusion of interest among men and women of different disciplines.

This was then the state of affairs at the beginning of the 1960's. Most of these developments were still young, but all were under way. It was the virtue of the Honolulu meeting to bring them together, and to present--truly for the first time--a viable interdisciplinary area specialism in the Pacific islands. It has not happened again, for the two subsequent Pacific Science Congresses, in Tokyo and Canberra, were in their very different ways unproductive and verging on the disastrous. But the field has undoubtedly been established, and it has flourished mightily in the succeeding decade.

Geography and the Pacific in the 1960's

The sort of geography that has evolved in the Pacific, as in some other third world areas, has been somewhat divergent from the subject as it evolved in the industrial countries during the 1960's, and it is of interest to speculate on the reasons. But it will be simpler to do this if we first draw back a little, and note that the modern entry of geography into the region did not come about until the contemporary revolution of the discipline was under way. It could not do so, because the older methods of description based on differentiation were wholly inappropriate for use in this environment, as a little consideration of some of the efforts in this area will clearly demonstrate.

Two or three attempts have been made to describe the geography of the Pacific islands in the classic regional manner: much the most noteworthy of these is Cumberland's.[24] The organization of the island sections of this book

---

pp. 362-382; "Les atolls océaniens: essai d'agronomie," Etudes d'Outre-mer Marseille, Vol. 40 (1957), pp. 252-267; Subsistence Agriculture in Melanesia (Honolulu: Bishop Museum, 1958); "L'agriculture polynésienne au contact des étrangers," Journal de la Société des Océanistes, Vol. 15 (1959), pp. 147-164; Subsistence Agriculture in Polynesia and Micronesia (Honolulu: Bishop Museum, 1961); "Histoire et préhistoire horticoles de l'Océanie tropicale," Journal de la Société des Océanistes, Vol. 21 (1965), pp. 55-78; "L'humide et le sec: An Essay on Ethnobiological Adaptation to Contrasted Environments in the Indo-Pacific Area," Journal of the Polynesian Society, Vol. 74 (1965), pp. 329-346; J. Barrau (ed.), Plants and Migrations of the Pacific Peoples (Honolulu: Bishop Museum, 1964).

[24] K. B. Cumberland, Southwest Pacific (Christchurch: Whitcombe and Tombs, 1954).

is interesting, particularly by the speed with which it has become a dangerous anachronism: Cumberland divided the islands into New Zealand's Pacific Neighborhood and Australia's Pacific Neighborhood. The island section of this book is substantial, and in the areas known to Cumberland the description is quite detailed. But it is difficult or impossible to achieve any sort of integration or balance by this method, and while there is no inherent reason why Rarotonga should not receive two pages and Sydney, N.S.W. two lines, it seems rather inappropriate in the absence of a strong thematic organization on island problems.

The island-by-island, or territory-by-territory approach meets with the serious problem that scale, data, and even basic descriptive information are widely unequal. The same problem arises in collections outside geography: thus a recent survey of land tenure problems, organized in this way, seeks to generalize the complexities of New Guinea's several hundred distinct societies in an equal and parallel chapter with the relative simplicities of, for example, Tonga.[25] It is partly for this reason that most of those who have sought to generalize widely over the Pacific have preferred to adopt a thematic organization of their material, which they have done with varying degrees of description and of analysis, and with varying success.[26] But for research purposes it was virtually impossible for serious geographical inquiry to begin until the modern liberation of the discipline. Land-use survey, once its technical problems of adaptation to the agricultural systems were solved, could only be an adjunct to other research: the problems and questions to which it led were, as several of us found, quite clearly "not geography" in the sense still generally understood as late as 1960.[27]

---

[25]R. G. Crocombe (ed.), Land Tenure in the Pacific (Melbourne: Oxford University Press, 1971).

[26]In particular, F. Doumenge, L'Homme dans le Pacifique Sud: étude géographique (Paris: Musée de l'Homme, 1966); H. C. Brookfield with D. Hart, Melanesia: A Geographical Interpretation of an Island World (London: Methuen, 1971); R. G. Ward (ed.), Man in the Pacific Islands: Essays on Geographical Change in the Pacific Islands (Oxford: Clarendon Press, 1972).

[27]I recall a sense of some despair when reading Hartshorne's Perspective on the Nature of Geography in the field in Chimbu in 1959-60, and drawing the conclusion that most of what I was doing seemed quite clearly to lie outside my own discipline. Later, in 1962 when giving a seminar on the relation of behavior in land tenure to land use systems, at a college of the University of London, I was told that since I was concerned with process and not with differentiation, I was moving outside the discipline. My reply was that if being a geographer were thus to restrict me, I would prefer to move outside. In all fairness I should add

Others found the same. Anthropologists faced questions beyond their own discipline. So did economists, political scientists and even prehistorians. The very scale of work forced upon us in a region of small countries, small islands, small societies, demanded that we probe in depth. And so we found ourselves talking much more readily with members of other disciplines united by the same set of research problems than with many of our fellow geographers. A few dropped out of geography altogether. Others decided to argue the toss with our colleagues, and to make a bid for the acceptance of what we were doing within our own discipline. I was among the latter, and a series of papers and later books has set out to argue a case which I do not intend to repeat here.[28] But this was not the whole story. We also incorporated some of the methods and theories of the "new geography," in an effort to disentangle the general from the specific. Thus I have explored distance decay in Chimbu agriculture in various ways;[29] Waddell has examined the location of settlement in relation to gardens in the Enga country of New Guinea, while Bedford has elaborately tested the gravity model and related models in studying New Hebridean internal migration, and has utilized and developed methods for deriving statistics from the graphs formed by the experience of individual migrants.[30] But in all these instances we have found that such hypothesis-testing has "explained" only a small part of the total question. The larger part of our interpretations have rested

---

that I was deliberately trailing my coat, and that I doubt if my critic on that occasion would say the same thing today. However, there remain those of us who seem to say that only what is "spatial" is geography.

[28] H. C. Brookfield, "Local Study and Comparative Method: An Example from Central New Guinea," Annals of the Association of American Geographers, Vol. 52 (1962), pp. 242-254; "Questions on the Human Frontiers of Geography," Economic Geography, Vol. 40 (1964), pp. 283-303; "New Directions in the Study of Agricultural Systems in Tropical Areas," in E. Drake (ed.), Evolution and Environment (New Haven: Yale University Press, 1968), pp. 413-439; H. C. Brookfield and D. Hart, op. cit. (1971).

[29] H. C. Brookfield, "The Money that Grows on Trees: The Consequences of an Innovation within a Man-Environment System," Australian Geographical Studies, Vol. 6 (1968), pp. 97-119; "Full Circle in Chimbu: A Study of Trends and Cycles," in H. C. Brookfield (ed.), The Pacific in Transition (London: Arnold, 1973), in press.

[30] E. Waddell, The Mound Builders: Agricultural Practices, Environment and Society in the Central Highlands of New Guinea (Seattle: University of Washington Press, 1972); R. D. Bedford, Mobility in Transition: An Analysis of Population Movement in the New Hebrides (Ph.D. thesis, A.N.U., Canberra, 1972, unpublished).

on data collected in depth and analysed painstakingly, leading to theory of limited domain generated by induction. Such theories have then become the ground for comparative testing in other inquiries. The search has more and more been for depth; the restrictions on data collection imposed by our labelling as "geographers" have been less and less observed; our work has therefore blended into an increasingly interdisciplinary thrust along with that of a steadily widening range of other social and natural scientists.

Stocktaking around 1970

Toward the end of the 1960's geographical research in the island Pacific began to move to the stage of fairly large scale synthesis over the accumulated results of inquiry, aided by a growing availability of data and the evolution of certain quite distinct lines of theory. First of these was that of Doumenge[31] and it has been followed by my three contributions[32] and by that of R. G. Ward[33] within geography, while in neighboring disciplines there have been overviews in prehistory, in the field of land tenure, and in the area of economic development --thus far only at article length. Books on more specific topics are growing more exploratory and comparative. A new *Journal of Pacific History*, founded in Canberra in 1966, has already become a major forum in a sense never achieved by the older *Journal of the Polynesian Society* and *Oceania*; the major geographical innovation, *Pacific Viewpoint*, has always had its eye on eastern Asia as much as on the Pacific, and its involvement with the island Pacific has been variable through time.

Since my own contribution features large in this list, I cannot discuss it in detail, but it is evident that work has moved to an altogether different plane, at which the accumulated results of two decades of mounting research effort have permitted a level of generalization that is regional in content, but also theoretical and more generic in scope. Within a context of interdisciplinary Pacific studies, Pacific geography is now able to make a very fair bid for its place in

---

[31] F. Doumenge, *op. cit.* (1966).

[32] H. C. Brookfield with D. Hart, *op. cit.* (1971); H. C. Brookfield, *Colonialism, Development and Independence: The Case of the Melanesian Islands in the South Pacific* (Cambridge: University Press, 1972); H. C. Brookfield (ed.), *The Pacific in Transition: Geographical Perspectives on Adaptation and Change* (London: Arnold, 1973).

[33] R. G. Ward, *op. cit.* (1972).

this area field: there are, indeed, not many of the world's recognized regions of area specialism in which the place of a well-diversified geography is so strong. The reasons, I think, are sufficiently clear. First we have contributed a large body of basic data, gathered both in general and in depth, which is of value to others. Second, we have been in the forefront in working on problems on the edges of two or more disciplines, and the work has been good enough, and innovative enough, to command respect from our colleagues in other disciplines. This is most widely true in cultural ecology, but also in population studies and more recently in the field of economic development. There are some outstanding examples in these areas, but I would isolate only three for special mention: Pirie's increasingly wide-ranging work on Pacific population geography, which has made him a leader in the field of Pacific population studies; Hughes' exhaustive and path-breaking study of trading systems in central New Guinea in the immediate pre-contact period, rigorous alike in its use of history, ethnohistory, and archaeology as well as geography; Clarke's elegant development of the ecological method, combining both American and Australian approaches to great effect, and leading on to his still-unpublished work on the historical and continuing place of taro in the cultivation systems of the whole of central and peripheral montane New Guinea. Collectively, these and several others range from within geography sufficiently deeply into demography, anthropology, ethnobotany, prehistory, and history as to be able to participate equally in these neighbor disciplines. One could develop this line of discussion in the form of an extensive annotated bibliography, but I have already been invidious enough in my selection, and refrain.

But we are also geographers, and must honestly face the fact that our impact on our fellows within our own discipline has not been of the same order. Indeed, to judge from much of the literature, we are known mainly for our few general theoretical or methodological statements, and hardly at all for the substantive work on which this froth rests. Yet *schaum ist kein bier* and it is the strong stuff beneath of which we are truly proud. Though this is a common fate of geographers--and perhaps one reason why, in desperation, so many of us turn to theory and method in order to get our message across--I believe it is an unusually marked feature with the Pacific group. The reasons why this might be so are worth closer examination.

The Dilemma of the Area Specialist

A generation or so ago it was often remarked that members of discipline "A" knew only their own literature, and never read that of discipline "B." This was the sort of situation which, with us, gave rise to what Carl Sauer once called the "pernicious anaemia" of the "but-is-it-geography" witch hunts, or which led Peter Lawrence in a delightful and salacious allegory[34] to pour scorn on a generation of anthropologists who shunned the use of documents and would have no truck with historical evidence. It is the sort of view that perhaps reached its apogee of statement--even if way beyond the optimal time for such a statement--in Gluckman and Devons' conclusion that in order to reach a "balanced" structure of theory within one discipline it was necessary to accept "limits of naivety" in relation to all other disciplines.[35] Today the situation has a different slant. Most human geographers no longer read any of the work of physical geographers; urban geographers never read the work of cultural geographers, and so on. We now read extensively in the disciplines related to our own field of geography, but a great many geographers read mighty little else written within their own subject.

This is not altogether bad, for as Harvey remarks there is a growing pressure toward integration in science, and especially social science, as a whole, and it is irrelevant if geography should disappear or disintegrate in that process:[36] Many intra- and inter-disciplinary boundaries have long since outlived their usefulness. But one consequence of the transgression of disciplinary boundaries is the evolution of area studies, and there is a real tendency for area literature to become the ambit of reading of a great many people; for many of them, it is the reading public of the same area literature that becomes the target of most or all of their writing.

I have spent a large part of this paper reviewing the very rapid growth of a distinctive field of area specialism in Pacific island studies since the second world war, and the important role that geographers have played in this creation. The dominant tradition in the social sciences in the Pacific was that of anthro-

---

[34] P. Lawrence, Don Juan in Melanesia (St. Lucia: Queensland University Press, 1967).

[35] M. Gluckman and E. Devons, Closed Systems and Open Minds (Manchester: University Press, 1964).

[36] D. Harvey, Explanation in Geography (London: Arnold, 1969).

pology, and it was in the Pacific that anthropologists evolved their particular research method based on the study of a small population in great depth. The rest of us absorbed this tradition as soon as we became area specialists in the Pacific, and in any case we were drawn into use of the same research method for the same reasons. In time we have sought and found ways recapturing a wider view, whether by use of theory derived from elsewhere, by development of the comparative method common to all the sciences, or more recently by essentially indigenous concentration on networks and flows fusing the local units into wider systems. The very fact of being on a research frontier within our area was exciting, and encouraging to innovation. We believed and believe that we were opening up new perspectives that would be valuable to geographers elsewhere, but it is only rather recently that we tired of waiting for them to find us by themselves, and set out rather consciously to sell the product.

It is because of the difference in scale that this appears to us a problem, while to the geographers of developed regions who use immense bodies of data all ready for the computer, no such problem seems to arise. Contrasted methods arose in different contexts, and evolved in different company. But there remains the common bond of geography, and whether or not it is irrelevant that this should in the long run disappear, it is as geographers that we make our contribution to area studies, and as geographers that we are there known.

The situation I am describing is relatively new. It stimulates efforts to seek and find common ground. Quite a few quantifiers of economic-urban background have now worked in parts of the tropics--some of them in the Pacific--and have faced for themselves the problems of adapting their ideas and methodology to a radically different situation. Similarly, some of us have moved our work into the developed countries: Waddell, for example, is adapting his methodology and concepts to the study of rural and maritime populations around the Gulf of St. Lawrence. Initial adaptation is not easy: my criticism of the "geography of modernization" concerns mainly its uncritical acceptance of an outworn diffusionist view of development and change.[37] But these are criticisms of an initial phase. Signs are not wanting of greater mutual understanding and greater willingness to comprehend the reasons for selection of a particular methodology. Given the fullness of time, and given a willingness to read at least those parts of the writings of different groups which are aimed at a more

---

[37]H. C. Brookfield, "On One Geography and a Third World," Transactions, Institute of British Geographers, Vol. 58 (1973), pp. 1-20.

than area-specialist audience, there is good reason to hope for a mutual benefit from what, at times, may yet be an acrimonious exchange. But the fullness of time may not be available to us, and this brings me to my most serious topic.

The Geographer and the Third World

No less marked than the gulf that has separated the area specialist from the alleged generalist who is in fact an area specialist in the problems of the western industrial world, is the gulf that separates the specialists of different areas which have a great deal in common. This gap is far less excusable, though there are obvious logistic reasons for its existence. Few Pacific specialists have much more than a random acquaintance with the literature of Africa, Latin America, the Caribbean, or even southeast Asia. And the converse is equally true. Nor is this surprising, given that a large proportion of this writing presupposes some acquaintance not only with the basic descriptive literature on each area, but also of its current economy, politics, and even personalities. We see articles and books of deliberately general intention, when these are published in places and in forms designed to bring them to wider notice, but the great mass of the regional literature remains hidden from sight, and the loss is very great for there are common problems which we all face.

Geography is a discipline centered in the developed countries of North America and Europe. The bulk of its literature concerns these countries, and most of its more international journals and publishing houses are based there. This is in contrast to anthropology, which is also based in the developed countries, but whose subject matter lies overwhelmingly in the developing world. As a profession, therefore, anthropologists tend to be more completely than any other aware of the new surge of nationalist xenophobia in the developing countries; they find themselves hoist with three generations of close linkage with the colonial and neo-colonial systems, even though they may have opposed colonial politics. Wriggle, and call themselves sociologists as they might, they are finding their close link with the whole pattern of rule, exhortation, and exploitation of the developing by the developed an immense handicap. Malinowski, whom we quoted above, was wrong in believing that there would soon be no "primitive" peoples for anthropologists to study; but anthropologists now find themselves in greater and greater danger of being unable to practice their highly-evolved craft, for reasons that very few of them could have foreseen.

Geographers face precisely the same risk in their participation in area studies. Thus far, we have only a few hints here and there in the literature, but it is already clear that a great many of us are going to be very sorry for some seemingly patronizing remarks we lightly put into print only a few short years ago. Our continuing value lies in the generality of our method and theory; where we are involved in foreign area studies, on the other hand, we partake fully of the colonial image. This is a problem that is still growing, and growing fast: it will not die away again in our generation.

Though xenophobia, like jingoism, and belief in so-called free-enterprise or communist systems, does not fall within the class of rational conclusion amenable to logical argument, we as a whole collective of expatriate scientists are being asked what our work has done to bring any benefit to the people whom we have been studying. A colleague, now living in Fiji, recently wrote to me of "academic hijacking" which--unlike the more normal form of hijacking--is socially acceptable, and brings both honor and rewards. The "academic hijacker" is one who obtains his grant to work in a developing country, does splendid work there, publishes it in London or New York, gains kudos, advancement, good offers, and a good life. He might send a few token copies of his publications to the people with whom he worked, but this is all the benefit they ever see. His teaching goes to his own students, who then wish not unreasonably to replicate so profitable a pattern. Do not ask whom he is describing: he is describing us.

Let us honestly face the fact that the sort of "area studies" we are discussing in a symposium such as this constitutes, in its construction, a fundamentally colonial club. Pacific studies, as yet almost the least affected by the hostile trends I am describing, is almost a classic example of such a club. Few of our colleagues are drawn from the countries we are studying; very few of our graduate students are drawn from these countries--and our grant-giving bodies are less and less willing to see the number even maintained, let alone increased. We go "out" to the field, and we try hard to involve ourselves there. We may even gain some real success in seeing the world through the eyes of the people, in gaining a "view from within." But we are only sojourners; to do our work we need to retire to base, to our libraries and colleagues, to the company of our fellow disciplinarians, and our fellow area specialists. Even though we may find a few centers where we can do this within our "areas," they are small and poorly served by comparison with the far wealthier and better supplied centers and networks of centers in the developed countries. Reasonably, because this is where the people are and the collective travel costs are less, a conference such as the present is held in Chicago, yet none of us is discussing North America. And if we were to locate ourselves for a few days somewhere in one of our "areas," in which one would we sensibly choose to be? And of what benefit would it be to move to, say Singapore, Kingston, or Ibadan, if we were all still a bunch of first-world expatriates as removed from the place and people around as were the "socially responsible geographers" of April 1971 from the city of Boston?

What I am saying is that a review of the role of geography in foreign area studies is likely to remain an essentially historical exercise--as I have made the bulk of my paper--unless the whole concept of area studies is radically revised. This is not to say that foreign area studies within developed countries are not necessary and even essential to the intellectual well-being of our own nations. But their role is in fact directed primarily to these nations: the wider value of area studies is quite inadequately developed.

Toward a Future

As I see it, area studies in their present form are a phase, a response to the convergence among the social sciences on the one hand, and to the massive growth of interest in the problems of development and change on the other. They are essentially a new form of compartmentalization of work still based, as ever, in the developed countries, and in which local institutions rarely have more than a secondary role. In an academic context they parallel the structure of international aid and technical assistance organizations, and even of the multinational corporations and conglomerates.

Thus in the Pacific the dominant centers of area study are in Canberra and Honolulu; secondary centers are located in New Zealand, Japan, and on the west coast of North America. Within the region are the universities of Port Moresby, Guam, and Suva; a very effective research unit is attached to the former, but is controlled from Canberra. There are also the very active offices of ORSTOM in Nouméa and Papeete--the latter including several geographers and doing some excellent work[38]--which are controlled from Paris. The overwhelming bulk of research capacity in the region is thus not only financed and staffed, but also controlled from abroad.

This is not an indefinitely tolerable situation, though it was inevitable as a stage in the growth of research capability. It must give way to a new sort of

---

[38]E.g., J. Fages, F. Revault, C. Rignon, and C. Robineau, "Tahiti et Moorea: études sur la société l'économie et l'utilisation de l'espace," _Travaux et Documents de l'O.R.S.T.O.M._, Vol. 4 (Paris: O.R.S.T.O.M., 1970).

structure, in which area studies *per se* become more truly localized. The appropriate centers for Asian studies are in Asia, of Pacific studies in the Pacific. But this is not to say that the "external" centers should vanish; rather they should utilize the most "underdeveloped" of their resources, which is the propinquity of scholars and of library materials covering a range of "areas." What is needed is a set of institutions--perhaps smaller, more concentrated, and more effective than the present plethora of so-called "area centers" and "area programs"--which are concerned with comparative work, and have truly operative linkages with the localized area centers. New sets of ideas are needed, and such comparative centers might for example be based on cross-disciplinary integration around themes such as "development studies"; the history of world interconnection, rather than colonial history; the classical arts of all lands; the ecology and economics of non-renewable and slowly-renewable resources. They need to move quickly, and far, from the outworn "area" concept.

In this way we might avoid the twin pitfalls of colonially-conceived area studies, and of area and disciplinary blinkers which limit the power of understanding. The institutional forms of the last thirty years have been of great value, but they have run their course; foundations, and government granting bodies have perhaps recognized this trend toward diminishing returns sooner than some of us, and we need new approaches if we are to recapture their imagination as well as ours.

But my argument does not rest only on pragmatic grounds. I left a secure and commanding position in one of the most strongly-established area-studies centers anywhere largely because of growing dissatisfaction with area blinkers. We drew in specialists from other areas, but all worked in the Pacific and East Asia. In its earlier, expansive days, the Research School of Pacific Studies spoke of its ambit as running from Madagascar to the Aleutians and Cape Horn, though we had a very proper concentration on the narrower region between Malaya and Samoa and this was our great strength. But I felt it was also becoming a weakness, and sought to test the earlier ambitions; I was not encouraged. So I came to a North America in which I hoped to find more catholic views--and have been grievously disappointed. Caribbean specialists in my present university are numerous, but most of them know little or nothing even of Latin America; Africanists everywhere in the continent see the developing world by a perception routed through Ibadan, or Nairobi; in a continent where national chauvinism is rising fast, and any real awareness of the world outside U.S.A., or Canada--and I say "or" rather than "and"--is simply fading away, the problem seems rather a fight to hold on to what we can of what we have in the "foreign area" field than any search for new comparative vision. The effort of trying to maintain one's place on a descending moving staircase is not conducive to innovation.

Yet even my own limited experience suggests something of the wealth of insight that can be gained from a comparative view. I offer a single, small example. An immense amount of research and writing has now been done on the alienation of land and recruitment of labor in the Pacific islands, though rather less on the Pacific plantation. There is an even greater quantity of writing on the slave plantation economy of the Caribbean, and the subsequent survival of the plantation by various means. In recent years, there is a rapidly increasing volume of detailed work on the evolution and continuity down to the present day of the *hacienda* system in Latin America. All these systems have in common the purpose of combining land and labor for production in a fundamentally capitalist economy; all have--in Marxist terms--necessitated the separation of labor from the "objective conditions" of its own productive activity; each evolved under different political, demographic, and economic conditions in different environments, and each has managed to persist as an ongoing system with greater or lesser modification through all subsequent changes; all have a common result in what Latin American writers term "marginalization" of the indigenous or resettled populations affected by the several modes employed to secure commercial production from the land. Yet there is almost no comparative discussion in the literature, and almost no attempt to seek illumination of regional processes by means of insights drawn from the processes described in other areas. Comparison between the evolution of 19th century and 20th century plantations under closely similar conditions of access to local and foreign labor is obvious enough, but it has rarely been done. Beyond this, one can see incipient *hacienda*-like forms even in the modern Pacific, and the reasons for their failure invite comparative consideration of the extraordinary success of such forms as adaptable socio-economic organisms in Latin America. And there are linkages. So tight was the control over labor by the *hacienda* in 19th century Peru, that new enterprises in that country initiated the overseas recruitment of labor from the Pacific islands and introduced a pattern that remained operative long after the Peruvians were excluded.[39]

---

[39] It would be inappropriate to offer even a selective bibliography covering these matters. All that is presented here are some very preliminary generalizations from work which I intend to pursue in much greater empirical and theoretical depth in future writing.

Conclusion

I return to the questions with which I began. Utilizing the Pacific as my example, I have shown how a modern "new" geography of quite variant form from the "new" geography much nurtured in this city has been able to make a very effective contribution in the rapid evolution of an interdisciplinary area specialism, and in so doing has found itself able to contribute a small share to the evolution of method and theory in our discipline. But I am far less sanguine in answering my second question: we have made little contribution of direct significance for the people among whom we have worked, and our efforts to achieve relevance have at best been partial, because our base is so firmly overseas. In consequence of remaining geographers, we have imported external models into our thinking the moment we have left the field, and begun the task of relating our work to the growing core of our subject. We have remained foreigners in thought, and our ambivalence between discipline and area has not been resolved.

Recently in New Guinea, the West Indian economist and activist Lloyd Best as written that any:

> arrangement demanding a strong center is . . . doomed to failure. Systems with strong centers generate charismatic leadership and demand large-scale political organization. Both of these are in conflict with native tradition and with the spirit of the place.[40]

Best is concerned with the structure of states, and also with the problem of regional development. What he writes is in conflict with almost all that we say in our discussions of modernization and development. By extension, it is also in conflict with the whole structure of foreign area studies, and their area-by-area link with disciplines whose core and thrust lies elsewhere. We, as geographers, should be able to comprehend the influences of spatial organization not only on development, but also on patterns of thought. It is time to draw a line under the achievement of foreign area studies, recognize the solid contribution that has been made, but recognize also the need for new patterns, and new spatial structures of inquiry.

---

[40] L. Best, "Self Respect and National Identity: Independent Thought in New States," Sixth Waigani Seminar, Port Moresby, 1972 (unpublished).

CHAPTER 5

GEOGRAPHIC STUDY OF CHINA

Rhoads Murphey
University of Michigan

I think it was Oscar Spate who once illustrated geography's position in area studies by a hypothetical statement of consensus among a group of scholars putting together a book on an area of which the first chapter was to deal with geography: "First geography, then let's get down to the important things." This may still be an accurate impression of how many scholars in other disciplines concerned with area studies regard geography; at least some geographers may have deserved some of this censure. It is unfortunately true that published materials by geographers which are relevant to area studies (as they all should be) and which are read and recognized by people in other area-oriented disciplines as valuable parts of the regional corpus are discouragingly few.

Area studies, like most important or exciting questions, are necessarily by their nature interdisciplinary. The imaginative scholar in any discipline quickly learns to keep his eye on the ball--to pursue the problem wherever it leads him and with what skills, attitudes, and tools he has or which he must acquire as the problem dictates, without worrying about disciplinary lines and disciplinary base. Even problems which come to his attention within his own discipline, as many will, or which seem wholly disciplinary and may begin as immediate or procedural needs will often lead him along the unbroken continuum of intellectual inquiry far from his starting place. He must also be continually aware of what other scholars are doing on the region of his concern, so that his own contributions can build on what already exists and can lead to new insights which will be valuable to others. The importance of the problem must also be considered, not how big or complex it may be but to what extent its pursuit may lead to the identification of other important problems or to the kinds of answers or insights which may have relevance beyond the original inquiry itself.

Importance of and Need for Areal Studies

The study of actual areas in their entire frame or gestalt would seem to be a (if not the) primary concern of the geographical discipline, as it attempts to study man-environment interrelations and spatial patterns. These life and blood matters of the discipline cannot be pursued with any validity except through the medium of real areas, nor can any area be properly understood without these basic dimensions. The geographer thus has a natural and fundamental role to play in area studies, and one which is intellectually challenging and exciting. Almost alone (together with some but by no means all historians) he attempts to see the region and its culture as a whole and to fit together all of its parts. As an anthropologist colleague of mine put it, "geography is the great integrating discipline." Yet a disproportionate share of the best minds in Western geography has for the past fifteen years or so been attracted into quantitative and theoretical work in geography which has so far been almost perversely non-regional --and, I would argue, as a result to a large degree non-geographic and non-real.

Most of those who work in geographic theory and measurement have not often enough tested or applied it in real regional contexts, even in North America or western Europe, or have done so inadequately, with little or no effort to understand the area concerned. Efforts have been concentrated instead on theory or technique, and as a result have often produced nonsense; they have not often enough pursued questions important to the understanding of the area concerned but only those which fit easily with pre-conceived theoretical objectives. In other cases, inadequate knowledge of the regional context has led to results of little or no validity or usefulness. Geographers whose work focusses primarily on an area have been given inadequate help by the model-builders. In fact, some of the most important and exciting applications of geographic theory to regional contexts have been made by non-geographers.[1] "Regional geography," a term which I tend to regard as redundant since all geography is properly regional, has tended to languish.

Certainly one reason is the demanding nature of the field. Sound area studies rest in most cases on a mastery of the area's language[2] and culture, of its scholarly literature, and of the work of scholars in a variety of other disci-

---

[1]See, for example, G. W. Skinner, "Marketing and Social Structure in Rural China," Journal of Asian Studies, Vol. 24 (1964-65), pp. 3-43, 195-228, and 363-399.

[2]On language training, see below, pp. 100-102.

plines concerned with the area. The acquisition of such mastery may well be the continuing work of a lifetime, as opposed to the acquisition of a technique. It is important if not essential also to develop at least a working familiarity with other areas, both similar and dissimilar, since without the comparative dimension in some depth, study of any one area will suffer seriously. It goes without saying that the regional geographer (if such a distinction has any meaning) must in addition have the same command of disciplinary and non-disciplinary research and analytical techniques as anyone else. This is a formidable set of requirements--and yet one would think that the rewards of area study were at least equally great.

The Case of China

It seems clear however that in the case of China they have not been perceived as great enough to produce a body of geographic study in any way commensurate with the size and importance of the area. China represents one of the two great poles of world civilization, and has for the past several millennia contained something like a quarter of the world's population. Both its traditional civilization and its current revolutionary experience offer models which contrast illuminatingly with experience in the rest of the world, as distinctive sets of answers to universal human problems. The geographical dimensions of all these matters are surely obvious. Yet the total of geographic literature on China, in all languages (and the attention devoted to it in university curricula) is proportionately tiny. If one is to be honest, much of what has been done by geographers is of only limited intellectual value and offers relatively little to other students of the area. The great bulk of it is purely descriptive, often elementary, often lacking in analysis, and often adding little or nothing to existing knowledge or presenting in simplified form information available elsewhere in the secondary literature. Given the paucity of detailed factual data (by modern Western standards) on China at almost any period, it seems notable that such a large proportion of the published literature deals with physical geography. This is true in particular of most of the German and Russian literature, and in turn of the Chinese, since Chinese geography as an academic discipline was modelled from the beginning primarily on Russian and German lines. Well over three-quarters of the articles in _Ti Li Hsüeh Pao_, for example, the leading Chinese geographical serial, have dealt with physical geography. The literature in English shows somewhat less emphasis on physical geography, and an approximately equal amount of attention to factual cataloguing and to simple description of small parts of the country, with periodic re-addressings of the tired old question of regionalization systems. There are of course many creditable and even distinguished exceptions, but on the whole they are exceptions.

It is not an exciting record of geographic contributions to the study of China. I have made a detailed search of the periodical literature since 1930, in Western languages and in Chinese, in some 47 geographical and interdisciplinary journals (journal titles are listed in an appendix to this paper), and sampled where I could several more, including a few Russian and Japanese serials by article titles; brief abstracts have been prepared for all articles which dealt in any way with China, Hong Kong, or Taiwan and which were written by geographers. The search yielded approximately 430 articles, about 200 of them in Chinese and about 30 in Japanese, although my search of the Japanese (like the Russian) literature was unfortunately far from complete. Setting aside for the moment the Chinese and Japanese literatures, the roughly 200 articles in Western languages offer a very mixed bag. Admittedly our present perspectives do not often deal kindly, whether or not justly, with most geographical literature as a whole for perhaps half of the period surveyed, namely 1930 to 1951, although in fact over three-quarters of the roughly 200 titles in Western languages date from the period since 1951. One is however tempted to guess that perhaps 50 of the 200 offer data, analysis, or ideas likely to be of value to other scholars. Although it would be gratuitous to single out examples, this relatively small corpus of useful work does include some outstanding contributions, imaginative, perceptive, and challenging, or breaking new ground in the collection, presentation, and analysis of data. For much of the rest, one is tempted to hope rather than to fear that other scholars will not read it. Too much of it is unimaginative cataloguing, classification, or description, and in far too many cases presents elementary information dealt with on a far more sophisticated level in the secondary literature by other scholars. In some of the worst cases it misinterprets or overlooks important implications of the matters with which it purports to deal, demonstrates unawareness of existing relevant scholarship, and suggests an approach which attaches little or no importance to intellectual analysis or to the work of other scholars.

One hears it said that geographers do not write many books (apart from textbooks), preferring to present research in article form. This does seem to have been the case so far with geographic work on China. There are nine or ten Western-language textbooks and/or handbooks on the geography of China and/or Taiwan in more or less current use. (This excludes, for example, earlier texts such as Louis Richard's A Comprehensive Geography of the Chinese Empire, published in 1908, but includes those of Y. F. Tuan, C. S. Chen, P. Gourou (in L'Asie), V. T. Zaichikov, P. M. Roxby, K. Buchanan, T. R. Tregear, G. B. Cressey, T. Shabad, and C. M. Hsieh.) There are many more textbooks in Chinese, of course, but they are primarily designed for secondary school use, as are most articles in the journal Ti Li Chih-shih (see appendix). Western-language textbooks are discussed below. The total of scholarly books on China by Western geographers is so far only about half that number--five or six altogether (not counting two books dealing solely with Hong Kong), depending on whether one regards certain authors (e.g., Walter Mallory, whom I have not counted) as geographers and how one distinguishes among monographs, photo-offset Ph. D. theses, and books. This literal handful of book-length research seems, whatever its quality, grossly out of scale with the size and importance of the field, although it must be acknowledged that the number especially of Western geographers professionally concerned with China has never been more than tiny. These very few book-length studies, all of them produced since 1950 and over half of them in the past three years, are however considerably more impressive in terms of their intellectual quality and of their usefulness to other scholars than the bulk of the periodical literature. One may hope that their number will continue to increase more rapidly.

The textbook scene seems less encouraging. It may be symptomatic of some of the problems of the discipline as a whole, and of that tiny part of it devoted to China, that there are nearly twice as many textbooks as scholarly books on the geography of China. Nor is it easy to admire them. As a sometime writer of textbooks myself I am in a poor position to criticize, although I am thus also painfully aware of the difficulties. As a teacher of undergraduate and graduate courses on the geography of China, however, I am frankly embarrassed to recommend to my students, or to scholars in other fields, any of the texts now extant. In some ways the field would be better off if none of them had ever been written, not only because their intellectual quality is generally so weak but because textbook writing is an enormous (and usually thankless) undertaking. Their authors might better have given an equivalent amount of time to research in this grossly neglected field and to the production of research monographs now so sadly lacking. Good textbooks must necessarily rest on a basis of extensive research and monographic output, ideally by both the textbooks' authors and by others in the field whose work can form the foundation for an effort at synthesis and summary. The field of Chinese geography offers no such corpus, the production of which must be given priority; until more is produced, textbooks are likely to remain inadequate.

Current and Persistent Difficulties

It is discouraging to those who offer university courses on the geography of China that course readings must still consist overwhelmingly of works by non-geographers, and that apologies must be made for so much of the geographic literature (including textbooks) as well as for its paucity. There is no question of the field's overriding importance, and one would have thought also its challenge, appeal, and excitement for geographers, of which more below. It is not easy to explain to students, or to one's self, why, if this is the case, there is so very little geographic research to point to, and why so much of what does exist is so unexciting. The obvious answer is no doubt the primary one: the extraordinary difficulty and additional time investment represented by the need to master the Chinese language and to control the diverse but essential literature of the field, primary and secondary, Chinese and foreign, in a spread of disciplines, as well as to maintain a continuing familiarity with the development of the geographical discipline and others closely related to it such as anthropology, economics, and history. In addition, the China field continues to suffer to a crippling extent from the virtual impossibility of field research and from the overwhelming lack of hard statistical data. Both have of course been especially true since 1949, but only slightly less so in the preceding decades. China is distant, difficult, or closed, and statistically arid. For any field work-oriented discipline or for one which relies heavily on quantitative data, it presents formidable obstacles. And yet exactly the same burdens lie on anthropologists, economists, historians, and political scientists. Admittedly their numbers and output in the China field are also relatively small, inexcusably small indeed, but nevertheless their work, simply in absolute let alone qualitative terms, dwarfs the geographic output. Even on a per capita basis the record of geographic research output on China is a very poor last. This is the less excusable because of the clear and logical concern of geography with the study of areas, and hence with the breeding of area specialists. Most of the other social sciences and even many of the humanities, as they are currently organized on American campuses, are slow to reward and quick to chide scholars who engage in area study or who become too much identified with work on a particular area. This should not be and never has been nearly as true for geography.

Most of the important and exciting work on China of geographic relevance has been done by non-geographers, such men as J. L. Buck, R. H. Tawney, Fei Hsiao-t'ung, Owen Lattimore, G. W. Skinner, Morton Fried, Joseph Rock, Joseph Needham, Ho P'ing-ti, Dwight Perkins, Albert Feuerwerker, H. B. Morse, K. C. Chang, Mark Elvin, D. F. Lach, C. P. Fitzgerald, Ramon Myers, Evelyn Rawski, H. G. Creel, Robert Hartwell, and a host of others. Some of their research has been based on field work or on quantitative data and analysis, despite the difficulties, and all of it has drawn on the extensive primary and secondary literature in Chinese, Japanese, and Western languages. It is not necessary, even for significant geographic research on China, to depend on field work. The discipline has in any case grown away from its earlier more central preoccupation with field study, and a large part of geographic research in the past ten or fifteen years has not been based on field observation.

The study of China indeed suffers in one sense not from too little source material but from too much. This longest enduring civilized tradition in the history of the world has also produced and preserved by far the largest body of records and other materials on its own experience. Almost none of it has been used by geographers, and yet huge chunks of it are directly relevant to geographic study. In addition to the massive dynastic histories there is the even larger mass of local gazetteers, prime geographic fodder as yet largely unsampled. Foreign contacts with China, beginning with Marco Polo's fascinating encounter (also ignored for the most part by geographers, *mirabile dictu*) and accelerating after the sixteenth century, and the development of Western and Japanese scholarship since the nineteenth century have also produced an immense body of both primary and secondary material, most of which has also lain unused by geographers. While it is not easy, largely as a result of language and culture differences, to work with some of the Chinese materials, it is perfectly possible, as the work of scholars in other disciplines demonstrates. There are also extensive English-language sources and data compilations, for example the statistics and Reports of the Maritime Customs Service and the voluminous British and American Consular Reports.

Language mastery may be over-emphasized in the training of area scholars. For the study of most problems in any discipline, it is not necessary to learn *all* techniques; specific techniques may be acquired selectively as the need for them becomes apparent. This may be applied to language, or to aspects of it, in a way which would be preferable to the often rigid and overwhelming language requirements which help to keep prospective scholars out of area studies. In any case, useful work can certainly be done in area studies without language mastery, given a knowledge of the area and its culture and familiarity with the Western literature.

I would be inclined to guess in fact that geographic work on China, especially before 1949, is at least as viable in terms of sources as on Latin America, Russia, or the Middle East, and far more so than on Africa, Oceania, Southeast Asia, South Asia before 1800, pre-Tokugawa Japan, or pre-seventeenth century North America. It is honestly difficult to understand why the geographic study of China has been so neglected for, as already suggested, its appeal, excitement, and importance seem so obvious. As it is, there are perhaps a baker's dozen of professional specialists active in research on the geography of China in North America, perhaps half that number in Europe, and not many more in the rest of the world outside China. Compare this with the total of geographers who work with Latin America, for example. The contrast, including the size and importance of the two areas, would be ludicrous if it were not also so serious.

Admittedly it is from the past two decades that one tends to expect the most acceptable output of geographic research in general, as well perhaps as the largest amount, and it is precisely those decades which have seen China closed to most direct observation--and many minds especially in the United States closed to the very existence of China. In this respect one may regard the parlous state of geographic research on China as one of the many tragic casualties of the McCarthy era in this country, a deadening hand still not entirely gone. Irrational anti-communism combined with inherent provincialism and naivete to virtually destroy the development of American awareness let alone scholarly attention to one quarter of the world, a quarter which already promises to play a disproportionate role in the landscape of the coming century, as it did for so long in the pre-modern past. The United States is at long last beginning to acknowledge reality in these terms, but the geographic profession, which ought to be in the forefront of such a long overdue change, lags very far behind even the man in the street, or the quite different men in the White House.

No one pretends that scholarly study of contemporary China is easy. But convenience has never been an appropriate guide to sound or significant research, in any discipline or on any problem. Even given the acute shortage of reliable or detailed statistical data, it is possible and indeed essential to pursue a great many basically important questions of a geographic nature arising out of China's experience since 1949. That period has seen and continues to see radical departures from traditional Chinese patterns whose study can illuminate both our understanding of China, past and present, and our understanding of the so-called "developing" world as a whole. It has also produced a series of new Chinese experiments with solutions to universal human problems, innovations whose significance and relevance we are just beginning to recognize. Most of them have direct geographic relevance, as is perhaps only to be expected in a culture where man's interrelation with the environment has always occupied a prominent place in livelihood, cosmology, philosophy, art, planning, social organization, and statecraft. These intriguing new developments, many of them of enormous significance not only for the Chinese quarter of mankind but for the entire world, are reported and discussed in considerable detail in the Chinese press (much of which is regularly translated into English) and in the English-language publications issued weekly and monthly by the Chinese government. The total of this readily available literature is in fact immense; geographers have used almost none of it, apart from a tiny handful of people. There are a number of bibliographical and research guides to this mass of material on contemporary China, in English, and scholars in other disciplines use it extensively, including a great many who do not read Chinese or who read it only with great difficulty. Why have most geographers remained so out of it?

If it is not because of the insuperable difficulty of the field--as I am suggesting it cannot be--perhaps it is because most geographers have at least in this respect warped priorities. The neglect of China may be attributable to provincialism, understandable perhaps but not excusable on the part of scholars in a discipline which is inherently concerned with the study of areas and yet which so largely ignores the single largest sector of the human geographic landscape. The study of China is not an arcane pursuit. China represents a huge chunk of human experience, historically a uniquely rich one, currently a uniquely challenging one. Ample materials exist for its study, even or perhaps especially from a geographical point of view. It is past time for the discipline to re-order its priorities. Recent concern with universals, with the development of geographic theory, and with the building of models for analysis and for development will remain seriously warped and incomplete while it so largely ignores a quarter of the world--not a quarter which can easily be subsumed into some larger category, nor a quarter which can safely be ignored on grounds of "primitiveness," unknowableness, non-comparability, or on any other grounds. Geographers cannot study so nearly exclusively Western man and his native or colonial world, and speak or write with any validity about mankind and his real total world. The Chinese experience is simply too big, too important, and too relevant to be treated, if at all, as a left-handed, by-the-way afterthought.

Importance to Geography of Contemporary Trends in China

I do not want to suggest that the contemporary scene is the only or even the most important focus for the geographic study of China. Indeed I would argue that there is no validity in the often-used term "historical geography," for the same reason that I reject "regional geography." All geography must by its nature be historical, as all geography must by its nature be regional (or as all valid history and all valid area study must also be geographical); and as already discussed, all geographic study of areas must be comparative or cross-cultural as well as interdisciplinary. Materials for the geographic study of the Chinese past are almost embarrassingly rich. The four thousand years of Chinese history, the record of Chinese occupation and use of the East Asian landscape, must surely offer the greatest of all challenges and reward to geographic study.

But there may be merit in stressing the urgency as well as the pragmatic and intellectual importance of remedying geographic inattention to developments in China since 1949, where so many things cry out for study by geographers. I cannot attempt even to list all of them here. But I have been struck by a number of recent efforts by North American and British geographers to identify the chief concerns of this rapidly evolving discipline as a whole; they are virtually unanimous in grouping a set of topics which might almost constitute a score card of the major ongoing developments in contemporary China which I am appealing to geographers to notice and to study. Although I feel sure that none of them had China in mind, their collective listing of the major current concerns of the discipline runs roughly as follows:

Urbanism, over-urbanization, and the negative consequences of modern Western urbanization models;

Environmental degradation, urban/industrial pollution, its perception, and the need for both study and action;

Resource appraisals and resource use as key culture variables;

The dynamics, and the menace, of population growth;

Economic growth strategies, and the urgency of the challenge of poverty and backwardness;

The crucial role of agriculture and its growth, in all of the above;

The quest for social and spatial justice and equity;

The search for valid geographic theory, based on regularities and universals.

There could be no better living laboratory for the study of each of these problems than China of the past two decades. Under any circumstances, it should be unthinkable to pursue any of them without careful consideration of their evolution in the Chinese context. Let me say something in these terms about each, necessarily briefly but at the risk of elaborating the obvious; too often I find, alas, that what should be obvious about China is not so to many scholars.

The Maoist anti-urban bias is clear, even if not wholly consistent.[3] It has obvious historical roots, in the history of the Chinese Communist Party and its struggle for power before 1949 and in the modern history of a China dominated by treaty-port colonialists who controlled and shaped nearly all of its large cities. These cities were also the homes of the Chinese bourgeoisie; cities were reactionary in the past, potentially revisionist now and in the future, and alienating at all times. Their growth in China continues to be controlled by an unprecedented policy which limits their size and which channels new industrial investment disproportionately into new or smaller cities in previously remote or backward areas, or into rural communes, which are to be made industrially as self-sufficient as possible but without acquiring the morally corrupting and alienating qualities of big cities, nor their damaging effects on the environment. City-dwellers, especially white collar workers, must spend a month or more every year, whatever their status, in productive physical labor in the countryside, where they may regain correct values. The distinctions between mental and manual labor, city and countryside, "experts" or bureaucrats and peasants or workers are to be eliminated. The benefits and the experience of industrialization and modernization are to be diffused uniformly over the landscape and to all of the people, while the destructive, de-humanizing, corrupting aspects of over-concentration in cities are to be avoided.

Communist China has reversed traditional perceptions of the harmony between man and nature.[4] Man is now seen, in Marxist/Maoist terms, as the

---

[3]The reader is referred here for a more extensive discussion to my "City and Countryside as Ideological Issues: India and China," Comparative Studies in Society and History, Vol. 14 (1972), pp. 250-267. Contemporary Chinese views of the city, and of the natural world (see the following paragraph), form an illuminating contrast with those of the modern West--as well as some intriguing parallels with contemporary Western thinking--and in all these respects one can see the interplay between rhetoric and reality. There is also a provocative contrast here with Soviet anti-rural, anti-peasant, and pro-urban attitudes, as another Communist and centralist state has developed quite different perceptions.

[4]In this paragraph I attempt a telegraphic summary of some matters dealt with in greater detail in my "Man and Nature in China," Modern Asian Studies, Vol. 1 (1967), pp. 313-333; even there, however, Chinese attitudes and behavior

lord of creation and as bending nature to his will. The environment must be struggled with and forced into man's service. The obstacles and threats it has always posed must be overcome, and its resources must be tapped, by the united efforts of a now coherently organized socialist man. But the environment must also be protected against despoliation, and its productivity in man's service must be enhanced. Erosion must be (and is being) controlled by massive reforestation. Pollution must be (and is being) controlled by careful technological and industrial planning, including industrial dispersal, severe limitations on the use of gasoline engines, and proper treatment and disposal of industrial and household wastes. The Chinese have of course long recognized human wastes as a vital resource rather than dumping them into the water supply system, an important and rational model ignored, like so much of the Chinese experience, in the so-called "developed" West. China is already as pollution-conscious as the United States, and has the enormous advantage of developing this consciousness before industrialization, urbanization, or the internal combustion engine have grown beyond beginning stages. China may well, in this as in other respects, lead the way for the rest of the world.

China is also acutely conscious of the size of its population (even though its total, incredibly enough, seems not to be known with any precision even in Peking) and of the problems posed by its further growth. After a period of intellectually fascinating but bitter ideological dispute, the decision was taken in the late 1950's to limit that growth in the interests of welfare. The period since then illustrates the almost unbelievable success of this uniquely closely organized society in implementing policy, even one so universally difficult to apply as restrictions on what are after all individual decisions of such an intensely personal and private a nature as reproduction. Tantalizingly, it is not known, apparently not even in China, exactly what the dimensions of this success may be on a national scale, but the techniques are clear, impressive, and full of potential promise for the large parts of the rest of the world which face the same or greater demographic problems. On the basis of highly selective and haphazard but numerous observations, it seems possible if not probable that China may achieve something close to zero population growth (as Japan seems also be be doing) by or before the end of the 1970's. The importance of this experience and of the model it offers can hardly be overemphasized.

---

toward nature are necessarily oversimplified. In any case, even contemporary Chinese attitudes differ from those of the West and probably include considerable carryover from the traditional past.

The Maoist strategy for economic growth seems bizarre to many Westerners.[5] It is strictly egalitarian, anti-elitist, anti-expert, and based on what Mao calls "the mass line." It discounts the concept of GNP as misleading or useless and stresses real welfare for all the people as the only legitimate goal of economic growth. The mass line has in fact succeeded in calling forth the enthusiastic commitment of nearly all people, of both sexes and all ages, to the challenge and the accomplishment of economic growth; it engages their talents and their innovative contributions as well as their labor. Communist China has been able not merely to mobilize but to galvanize over 800 million people in the conquest of poverty. One of the world's poorest and unhealthiest countries twenty-five years ago, China has achieved economic security for all its people and has become one of the world's healthiest countries, with a public health system beginning to be recognized as a model not merely for the third world but for all countries. The human, aesthetic, intellectual, and philosophical costs of all these achievements are not denied, but this is not the place to evaluate them or to distinguish between Chinese and Western values and perceptions. The challenge of the Maoist strategy is that it works, and that it delivers economic, social, and spatial justice in a form which does indeed challenge the urban/industrial model of the West while at the same time offering both promise and appeal to the rest of the world.

Agriculture has played a key role in this transformation, although policy has been inconsistent over time and recurrent errors of judgment and planning have had to be corrected. The Chinese experience nevertheless demonstrates what can be accomplished, and how it can be accomplished, in this critical sector. Chinese agriculture was traditionally intensive, highly developed, and highly productive. In these terms, China is not really comparable to most of the rest of the world. But the success in approximately doubling agricultural production (again the quantitative data are lacking) by a combination of new technology, new investment, and new organization and intensification of labor inputs is impressive and offers a further model for application or modification elsewhere, at a time when agricultural development problems are urgently pressing. Chinese experience here contrasts especially with Soviet experience, both because it has been much more rapidly successful and because it has nevertheless permitted peasant consumption levels to rise almost as quickly and sub-

---

[5]Including Russians; among many other things, one may note the contrast between Maoist suspicion of the bureaucracy and the Leninist-Stalinist emphasis on Party structure.

stantially as urban levels rather than squeezing the peasant agricultural sector in order to extract capital for industrial investment to a degree which would result, as in the Soviet case, in deterioration or stagnation of rural incomes. This doubling of agricultural production in China has also been accomplished so far with minimal mechanization, a strategy suited to China's still proportionately small industrial sector and hence also more likely to be applicable in other third world situations.

Conclusion

In their organization of space, in their identification, appraisal, and allocation of resources, in their goals and strategies for development, and in their evolving solutions to a host of universal human problems too numerous to discuss here, the Chinese represent a part of the human experience, and of the geographic landscape, which cannot be ignored. The applicability of Chinese models to the development or management or reformist efforts of other areas may well be debated, if only because the Chinese system and society are so strikingly organized and controlled, drawing on a long tradition of close social integration, and so highly developed or sophisticated, again based on a long tradition of organizational and managerial sophistication. China is not and has never been a backward culture, and to that extent its experience and achievements may be only partly transferable to much of the rest of the world. But the Chinese experience is nevertheless enormously important as a model, both intellectually and practically--and I would say once more, especially for geographers. The sweeping themes touched on above have been dealt with at best tangentially by a handful of geographers, and not because attention has been given instead to the far larger canvas of traditional and pre-Communist China.

We appear to have crossed a watershed in relations between China and the rest of the world. After twenty years of virtually complete isolation, contacts are beginning again. I had the good fortune to return to China for nearly a month in May of 1972. I hope and believe that many more geographers, and many more Americans, may be able to do so in the next few years. Scholarly research in China is still not possible and may not be for some time, nor is free and full exchange of information, viewpoints, and publications. China is still in revolution. But the interchange of people and of ideas has begun and will increase, and the body of published materials for the study of China will continue to grow. Meanwhile, even structured and formal visits, such as mine was, and such as all visits are likely to be for some time, will continue to provide increasing opportunities for direct observation, much of which can be of tremendous value, both generally and specifically. China has never been an unknown or unknowable country--except, it would seem, to most geographers. The importance of the Chinese experience is becoming clearer all the time. It is not too late for the geographical profession to respond.

APPENDIX

List of Journals Searched for Literature on China, (1930-1972)

North American

- Annals of the Association of American Geographers
- Economic Geography
- The Geographical Review
- The Journal of Geography
- The Professional Geographer
- The Canadian Geographer
- The Canadian Geographical Journal
- Comparative Studies in Society and History
- Economic Development and Cultural Change
- The Journal of Asian Studies
- Pacific Affairs (Vancouver)

United Kingdom

- The Geographical Magazine
- The Geographical Journal
- The Scottish Geographical Magazine
- Modern Asian Studies
- Regional Studies (Oxford)
- Transactions, Inst. of British Geographers

Other English Language

- Proceedings, International Geographic Congress
- Contemporary China (Hong Kong)
- Journal of Oriental Studies (Hong Kong)
- Journal of the North China Branch, Royal Asiatic Society (Shanghai)
- Bulletin of the School of Oriental and African Studies (London)
- Journal of the Economic and Social History of the Orient
- The Australian Geographer
- The New Zealand Geographer
- Journal, Hong Kong Branch, Royal Asiatic Society
- Pacific Viewpoint (Wellington, N.Z.)
- Journal of Tropical Geography (Singapore)
- The Oriental Geographer (Dacca)
- The Indian Geographical Journal (Madras)
- The Indian Geographer (Delhi)
- The Geographical Review of India (Calcutta)

Germany

Erde (Berlin)
Erdkunde (Bonn)
Geographische Zeitschrift (Wiesbaden)
Mitteilungen, Geog. Gesellschaft, München
Petermanns Geographische Mitteilungen (Gotha)
Zeitschrift für Erdkunde (Frankfurt a.M.)

France

Géographie, and successor Acta Geographica
Bulletin de d'Association de Géographes Français
Annales de Géographie

Sweden

Geografiska Annaler

Chinese

Ti Li Hsüeh Pao (Acta Geographica Sinica)*
Ti Li Chih-shih (Geographical Knowledge)
Ti Li (Geography) Peking, 1959-
Ti Li Hsüeh Tzu Liao (Materials for Geographic Study)
Ti Li Hsüeh Yen-chiu (Geographic Research) (Taipei, 1966-)

Japanese (incomplete sample, mainly by title only)

Chirigaku hyoron (Geographical Review) Tokyo
Jimbun Chiri (Human Geography) Kyoto
Chigaku zasshi (Journal of Geography) Tokyo
Rekishi chirigaku kiyo (Proceedings of Historical Geography) Tokyo

Russian (incomplete sample, almost wholly by title only)

Izvestiia. Seriia Geograficheskaia (Moscow)
Geograficheskoe obshchestvo. Izvestiia (Moscow)

---

*Begun in 1934, apparently suspended in 1966 with the Cultural Revolution, but difficult to obtain abroad after about 1959; most American holdings are very spotty and I have been able to consult less than half of the supposed issues since 1959.

CHAPTER 6

TOWARD A REDISCOVERY OF INDIA: THOUGHTS ON SOME NEGLECTED GEOGRAPHY

David E. Sopher
Syracuse University

In his preface to what has long been the standard South Asian geography, O. H. K. Spate notes that writers on Indian affairs form a special class of ruminants, chewing over, with ever diminishing returns, the cud of government reports and statistics.[1] One is led to wonder about the possible relationship between diet and physique, since geographic members of the breed, though not few in number, do display tell-tale signs of emaciation. Yet having grazed for some years on the Census of India, I am reluctant to accept so simple an explanation of this pathology; I would want to look more closely at the ecology of the geographic discipline in the Indian situation.

We are asked here to consider geographic work on South Asia and other regions in relation to what are seen in this country as "foreign area studies." This need not restrict discussion to the work of geographers who are not Indians, but this is what I shall want to emphasize. Such work continues to be a major source of innovation in the geographic study of India, urging the methods to be used and posing the questions that are pursued by most Indian geographers. One has not yet heard the question raised, as it has been in some other disciplines, as to whether there can be or ought to be an indigenous Indian geography. That there is in any case not now such a geography has some bearing on an explanation of what appears to be a case of stunted development.

When we examine the nutritional resources of geographers working on the Indian world, we are struck by what seems to be a paradox. However well chewed over some patches may be, accessible stocks of state materials--reports,

[1] O. H. K. Spate, India and Pakistan: A General and Regional Geography (London: Methuen, 1954), p. vii.

gazetteers, censuses, economic statistics, topographic maps, all suitable for the feeding of certain kinds of geographers--are extraordinarily rich. Some remain almost undisturbed. A geographer who is not an Indianist, Wilbur Zelinsky, has recently expressed his wonder, and not a little envy, at the extent of this treasure.[2] It is disappointing, then, to find that the yield in geographic work from this massive food resource appears to be mediocre. The answer would seem to lie not in a deficiency of caloric content, to flay this exhausted metaphor for the last time, but in a serious dietary imbalance.

Strangers, Officers, Dwellers

Consider the healthy geographies we know. We can see that these have grown out of the perceptions of three different kinds of observers: strangers, officers, and dwellers. The interplay of ideas and conceptualizations based on their different perceptions has long been present (and continues to be) within much of western geography. These categories are to be understood as roles; officers who record their observations as strangers will select material for observation, fitting it into their personal world-view and for their own purposes, that they might not while imposing a more rigid, more distant and impersonal view as officers. Not infrequently in India we find the distant official frame of observation being used by persons who happen to be dwellers, natives of the place they may subsequently observe as officers. To an unusual extent, geography in India has been based (and continues to be) on the compilation of observations made for the purposes of the state. Certainly, the respected tradition of the intelligent and inquiring stranger in India, that provides us with indispensable accounts of the country at intervals over a period of so many centuries, also found practitioners among British officers in the early decades of their rule in India--Wilks, Buchanan, Pottinger, Sleeman, many others. Such traveler's accounts, and accounts of Indians in parts of the country foreign to them, have long since been overwhelmed in quantity and in demand by the flood of state-generated and state-related data. Personal records, diaries, and travel journals, published and unpublished, have scarcely been touched in more recent decades by geographers in India, who themselves seem less and less inclined

---

[2]Wilbur Zelinsky, "India (or Non-India) Seen from Afar: On the Mutual Relevance of Varied Cultural Geographies," in Robert I. Crane (ed.), Regions and Regionalism in South Asian Studies: An Exploratory Study (Duke University Program in Comparative Studies on Southern Asia, Monograph No. 5, 1967), pp. 139-146.

to use their own observation for insights or for pointers to new questions, while eliciting the raw data meant to answer old ones. The record is not entirely bleak: one knows, for instance, of Fritz Bartz poking about the dunes and backwaters of the Indian strand,[3] and Joseph Schwartzberg pedaling his bicycle from village to village on the dust-shrouded plain of northern India.[4]

What is missing in the Indian geographic tradition is the dweller's perception, one that is present as an integral element of human geography in much of the West. This need not be the geography of places as the natives themselves conceive it, such as some current research is trying to recover. It should be a geography in which the popular landscape symbols, mental maps, and "sense of place" of the dwellers, that is, the folk ecosystemic models, are made to interact with and become incorporated in the intellectual constructs of strangers and officers. This is what we sometimes find in the best regional geographies of an earlier European generation. The local perceptions that these draw on are, certainly, often expressed in a cultivated, though not necessarily cosmopolitan, voice--as an interest in local history, antiquities, folk tales (which may richly inform local geography), place names, and topography.

One thinks of the French geographies of pays written early in this century; Vallaux's Basse-Bretagne and the Morvan of Levainville[5] exemplify particularly well the land as interpreted through technical studies of the physical environment, one's own tutored observations, and careful gleaning of the local lore, both written and oral. A work that might well find application in the Indian situation would be Gallois' study of the names of pays in the Paris Basin,[6] with its repeated injunction to geographers to ask the countryman, the dweller, what he knows of and how he conceives of the land he inhabits. In the Indian literature, almost

---

[3]Fritz Bartz, Fischer auf Ceylon, Bonner Geographische Abhandlungen, Heft 27 (Bonn, 1959); Die grossen Fischereiräume der Welt. Bd. II, Asien mit Einschluss der Sowjetunion (Wiesbaden: Franz Steiner, 1965).

[4]Joseph E. Schwartzberg, "A Strategy for South Asian Regional Research," in Robert I. Crane (ed.), Regions and Regionalism in South Asian Studies: An Exploratory Study (Duke University Program in Comparative Studies on Southern Asia, Monograph No. 5, 1967), pp. 112-138.

[5]Camille Vallaux, La Basse-Bretagne: étude de géographie humaine (Paris: Cornely, 1906); J. Levainville, Le Morvan: étude de géographie humaine (Paris: Armand Colin, 1909).

[6]L. Gallois, Régions naturelles et noms de pays (Paris: Armand Colin, 1908).

the only exemplar is Geddes' Au pays de Tagore[7] which is in the direct tradition of Vidal de la Blache, Demangeon, and Sion. Geddes stayed for several months at Shantiniketan where the poet Rabindranath Tagore maintained a school in the Bental countryside; he also had opportunities to travel about Bengal, and had contacts with leading Bengali intellectuals. His interest was to work out from evidence in the landscape and in local administrative records the history of ecological transformations and shifts in locational significance that the vicinity of Shantiniketan had sustained. While his main thesis (which has been questioned) is to demonstrate a causal connection between deforestation on the upland margins of the delta and siltation lower down, and a succession of causally linked changes in drainage, the ecology of malaria, regional productivity, and population density, his perceptive sketches of village culture and his efforts to decipher the social meaning of location and distribution of settlement features have rarely been emulated in Indian geography until quite recently. In the interim, we are obliged to turn to the spirited intelligence of a sociologist, Radhakamal Mukerjee,[8] for an exposition of Indian settlement geography that tries to take proper account of social processes.

Geddes' model of landscape geography has remained inaccessible to most Indian geographers, never having been translated into English, perhaps because its geographic idiom was also somewhat foreign to a predominantly British-trained profession. Of course, this particular idiom is now almost obsolete in the West, dismissed by many as archaic, nostalgically seeking an ancient past. Yet the tradition has been important in giving form to academic geography in the West, and one can conceive of extant but unwritten geographies that serve to keep a rein on current theorizing, maintaining the dialectic between scientific abstraction and the observable world, the one that we can see, hear, smell, and touch. While classic landscape geography may have failed to adjust its scale of investigation to technological changes in circulation, its potential usefulness for Indian studies is not thereby invalidated; as Pinchemel has recently pointed out once again, landscape is "the expression of a civilization."[9] In the virtual

---

[7] Arthur Geddes, Au pays de Tagore: la civilisation rurale de Bengale occidental et ses facteurs géographiques (Paris: Armand Colin, 1928).

[8] Radhakamal Mukerjee, The Regional Balance of Man (Madras, 1938); Man and His Habitation: A Study in Social Ecology (London: Longmans, 1940).

[9] Philippe Pinchemel, "Redécouvrir la géographie," Annales de l'Université de Paris, Vol. 38 (1968), pp. 350-360.

absence of such a tradition in Indian geography, it is almost impossible for us to construe the symbolic information communicated in the Indian landscape, as we should wish to do if Philip Wagner's recent exhortation[10] is to be attended.

Paradoxically, we discern a glimmer of feeling for a piece of Indian land as home, of that sense of place that is so prominent in genteel thought in the West in the sympathies of one or another British officer a century ago, recording impressions in the field and compiling local lore, with the help of munshi, pandit, and clerk, for the geographic handbooks known as District Gazetteers. Philip Mason conjures up an idealized picture of the District Officer at work: "And incidentally, being a fair scholar, a man who perhaps got a good second or just scraped a first, he is quietly getting on whenever he has a spare moment, with the *Gazetteer* describing the history, antiquities, crops, castes, sects, marriage customs, land tenures, soil, and occupations of his district."[11] These district gazetteers often suggested models of ecological and spatial relationships that were basic to ideas appearing in subsequent regional geographies. Statistics were frequently used, and when the officer knew his district he did not hesitate to qualify a misleading appearance that might be conveyed in Administrative Report or decennial Census by the use of inappropriate or ambiguous statistical categories. The practice, unfortunately, seems to have lapsed.

All this is not to say that a sense of place is missing in India, only reference to the cultivated expression of it in the work of geographers. As the field observer quickly learns, powerful symbolic images attached to place are to be found everywhere, although they may often be a matter of intimate knowledge. They are seen in the ritual marking of points and patterning of lines that structure living and working space and in the web of relationships that make up an individual's ritualized social field.[12] A vital folk geography remains largely covered over in the "state-oriented" geographic literature. Not surprisingly, it is Geddes who points to the existence of North Indian analogues of the *pays*, namely, *bhūm* and *deṣa* ("desh"), while a recent wide-ranging survey of regional concepts

---

[10] Philip L. Wagner, *Environments and Peoples* (Englewood Cliffs, N.J.: Prentice-Hall, 1972).

[11] Philip Woodruff Philip Mason, *The Men Who Ruled India*, Vol. II: *The Guardians* (London: Jonathan Cape, 1954), p. 182.

[12] For a discussion of the "topographical religiosity" manifested in Indian villages, see Kees W. Bolle, "Speaking of a Place," in Joseph M. Kitagawa and Charles H. Long (eds.), *Myths and Symbols: Studies in Honor of Mircea Eliade* (Chicago and London: The University of Chicago Press, 1969), pp. 127-140.

by a leading Indian geographer discusses French and German terminology, but says nothing about the common North Indian usages.[13]

Uncovering the folk geography of India at the local scale would certainly be complicated by the fact that mental maps, perceptions that are affected by experienced fields of social interaction, can be strongly differentiated for different caste groups in the same locality. The expectable absence of consensus may help to explain why a local sense of place is not often evident, and perhaps because of this, so infrequently elicited. And then, the special point of view, the symbolic ties to place, of a closed social group may be habitually kept guarded. An Indian geographer on home ground may prefer to play the stranger in his writing, while the stranger from abroad seeks to simulate the dweller's perceptions.

But even the accessible language of the material landscape, disclosing a rich repertory of symbols, is only sparsely recorded in the geographic literature of India. The deficiency becomes clear when one is in the field. Distinct landscapes--organized ecological system-spaces--can be made out; a brief enquiry may yield common local designations for these and other ecological unities of space not found on current maps and different from the universally employed administrative toponymy. Throngs at shrine and fair make manifest the spatial structure of rural life, as do, at certain seasons, village bridal parties in their finery, squatting patiently on the platforms of small-town railway stations. Neophyte from abroad or older hand, the geographer immersed in the local landscape feels dizzy with the sense of discovery: so much that he sees is new, not marked out in his reading by earlier guides in the discipline.

Assessment of Some Major Works

How now to review this reading, to assay the sprawling corpus of "the geography of South Asia"? Able to claim only a modest acquaintance with a small part of it, I propose to consider in some detail a handful of major works. These represent both a summation of a larger body of writing extant at a particular period and the distinctive point of view of a figure of some importance. My critique will be directed by hindsight: from an awareness of where we are now (and some uneasiness that we are not where I think we ought to be), I want to look at the *magna opera* I have selected, asking about the intentions of their

---

[13]R. L. Singh, *India: A Regional Geography* (Varanasi: National Geographic Society of India, 1971).

authors, their assumptions, their methods, the paths along which they have pointed, and what they may have had to say to those outside the confines of the geographic discipline.

Let me concede that my comments, on occasion, may appear to be narrowly directed at particular cases. I would argue that such apparently isolated cases are demonstrably common forms and practices or ones that have been well received, and widely, within the discipline. By trying to probe a few works, I must run the risk of appearing to generalize from single instances and even to be arguing contentiously ad hominem, but by doing so I can hope to expose something of the complicated structure of ideas that underlies the academic effort of geography.

It may also appear to be unfairly circumscribing the geographer's contribution to knowledge to exclude some of what, at one time or another, has been done in the name of geography, such as the difficult task, begun more than a century ago, of surveying India and mapping it on a large scale. Concern in this paper is with the academic discipline of geography, and the Survey of India's topographic maps, the district gazetteers I have mentioned, and the work of Arab and Indian precursors of the gazetteer compilers are to be considered as resources rather than products of academic geography and its intellectual forebears. Academic geography has of course made its contribution to this work over the past century, collaborating in the creation of these resources in an effort to improve their quality.[14]

These resources were already being put to use in 1883 in Elisée Reclus' L'Inde, which appeared in the eighth volume of his encyclopedic overview of the late nineteenth century world.[15] It is a work of distinction, displaying with assurance the spatial patterns of the major themes of Indian history and remarking the variety of the Indian world set within the larger one. (But Reclus' geopolitical forecasts, drawn from these observations, have proved, like so many of the kind, to be wide of the mark.)

While the book is in part based on the early gazetteers and the first comprehensive Indian census, that of 1881, it draws more from Indological studies of its period than does most subsequent geographical writing. The enormous distance between author and subject--Reclus never visited India--was thereby

---

[14] See, for example, F. J. Richards, "Human Geography and Administration in India," Geography, Vol. 14 (1927), pp. 46-48.

[15] Elisée Reclus, Nouvelle géographie universelle: la terre et les hommes, Vol. 8: L'Inde et l'Indochine (Paris, 1883).

bridged; by referring to a model of a specifically *Indian* civilization, even an admittedly archaic model, Reclus was able to fix in place important symbolic elements of the mythic map of living Indians. Thus, he presents the classical North Indian calendar, noting briefly how it is related to the meteorological realities, a theme that most later geographers will choose to ignore. He pays attention to places invested with sanctity and to their legends, and he maps them (as well as the location of Catholic and Protestant missions), again not to be followed in later regional syntheses.

What did Reclus have to offer the Indianist in return, remembering, of course, that the latter was not his chief intended audience? Although he presents a framework of "natural regions," he gives surprisingly little expression to that preoccupation of later geographers, the explanation of human activity in relation to physical environmental variables. Rather, his regions are a framework of what we now call "spatial organization," partly closed fields of human activity, that may be characterized, but need not be, by a high degree of uniformity. These are still prototype natural regions, loosely keyed to the notion of river basins; because they are not insistently imposed, they show themselves to be flexible and quite serviceable. There is no insistence on bounding lines, and regions simply shade off at a distance from the center. Yet the organization of the whole into separate regional chapters does make for a strong sense of closure, of a degree of compartmentalization that appears to deny the reality, as so much regionalizing does, of the continuum of scale levels at which areal clustering of like properties and values occurs. Thus, commonalities at the all-India "macro-regional" level are identified, while below there appears a mosaic of some convenient number of regional blocks. Ignored are the varying strengths of interregional connections and gradual change that may occur without a break at the "regional" boundaries while cumulatively generating large contrasts. The latter pattern, which might be a trend surface of low order, is exemplified in the conventional recognition of a "North-South" cultural gradient and a "wet-dry" ecological contrast.

Half a century after Reclus, the state of the art is demonstrated in a typical regional monograph of the time, Krebs' *Vorderindien und Ceylon*.[16] Set beside its contemporaries--such products of the same German orientation to *Landschaftskunde* as Credner's study of Siam, Kolb's of the Philippines, Lauten-

---

[16] Norbert Krebs, *Vorderindien und Ceylon: eine Landeskunde* (Stuttgart: J. Engelhorn, 1939).

sach's of Korea[17]--its peculiar defects, arising from the distance between author and subject, become conspicuous. Krebs was aware of this shortcoming, but could find little in the discipline's literature, and could himself only skim the resources of census and gazetteer, to enable him to overcome that distance. Attempting to see India as a civilization, he could spend only a few months in traveling by car (his itinerary is not given but is marked by his photographs of landscapes in widely scattered locations).[18] We sense Krebs' yearning to connect what he has read of culture and history to what he sees, but he is overwhelmed. He can offer only sketches of descriptions and explanation that bespeak careful tutoring, yet remain unsure. He gives us a view from the train window. Green gives way to brown, or basalt to gneiss, habitations coalesce and scatter again, coconut palms fill the frame here, palmyra there, and the familiar configurations of rice paddy, of fields of sugar cane or <u>arhar</u>, are recorded as they appear. But the human societies that lead their lives outside cannot be heard, although a dash of color--an ethnic label or other--may enliven the sketch. We can feel the poignancy of a mission failed when for once people do make a brief appearance--on the veranda of the German Club overlooking Bombay's Back Bay.

Krebs' failure was nevertheless a valiant one, given the deficiencies of geographical research in India then. In the decades since Reclus, relationships that were tentative for the latter had become well established: properties of soil, certain relationships of climate to soil and vegetation, to agriculture and population density, the outlines of recent earth history and, important in their implications for human prehistory and history, some suggestions as to surface alterations during previous decades, centuries, and even millennia. Krebs comes at this last theme time and again. His field experience of the cultural geography of both central and southern Europe makes him wary of the way in which forestry specialists and the occasional historian have attempted to reconstruct the pristine Indian vegetation and that of later periods but there has simply not been enough research for him to make an effective case. He cannot quite tie

---

[17] Wilhelm Credner, <u>Siam, das Land der Tai: eine Landeskunde auf Grund einiger Reisen und Forschungen</u> (Stuttgart: J. Engelhorn, 1935); Albert Kolb, <u>Die Philippinen</u> (Leipzig: K. F. Köhler, 1942); H. Lautensach, <u>Korea</u> (Leipzig: K. F. Köhler, 1945).

[18] The field work was done in 1931-32. Appearing at the outbreak of World War II, the book was to become a rarity within a short time; it was reprinted in 1965 by Wissenschaftliche Buchgesellschaft, Darmstadt.

together the grand theme of human modification of the Indian earth, and later geographers will only fitfully approach it.[19]

Despite the contemporary demand for precise and often elaborate regionalization, Krebs is unusual in offering only a sketch of culture "spheres"; for these, he uses a term that is not common, Bezirk, which conveys a sense of circulatory flow. There is a feeling for core and hinterland, and the descriptive modulations across zones of transition are conducted with skill. Benefitting from hindsight, we can see in his broad historical models of circulation perhaps too great a reliance on the diffusion of Muslim influences as explanation of certain cultural surfaces and perhaps too little awareness of the full cultural reach of the great port metropolises. Certainly the questions he asks or points to are significant for Indianist as well as geographer. Yet without an intimate understanding of the way society is organized, especially in its spatial dimensions, without the systemic understanding of relationships at a succession of scales, between land and people, community, society, the answers he would have wanted could not be provided.[20]

When we turn to Spate's remarkable work of regional geography, first published shortly after independence and partition of the Indian region, and substantially revised a few years ago with the collaboration of A. T. A. Learmonth, we meet an unusual intelligence with a particular point of view, using with extraordinary skill the materials at hand. There is little that matters even a little in the geographic literature of India that has not come to his notice and has not then been worked in somewhere in the grand composition, if only as the subject of a footnote or an ironic aside. The workmanship is admirable, as is the display of

---

[19] The gap was noticeable in the great symposium, held in 1953, whose proceedings are published in William L. Thomas, Jr. (ed.), Man's Role in Changing the Face of the Earth (Chicago: University of Chicago Press, 1956).

[20] For example, we may be taken aback today by his assertion that the traditional Indian village community formed a "closed unit" (eine geschlossene Einheit). But, then, it is an idea that dies hard in the geographic literature, and even the latest edition of Spate's book seems to accept the notion: no reference to the intricate pattern of social linkages among Indian villages appears to qualify the tired metaphor that speaks of them as "so many unfederated little republics" (O. H. K. Spate and A. T. A. Learmonth, India and Pakistan: A General and Regional Geography, 3rd rev. ed. [London: Methuen, 1967], p. 169). See in this connection David G. Mandelbaum, Society in India (Berkeley and Los Angeles: University of California Press, 1970), one vol. in 2, pp. 352-357; 381-404. Clive Dewey, "Images of the Village Community: A Study in Anglo-Indian Ideology," Modern Asian Studies, Vol. 6 (1972), pp. 291-328, reviews the shifts in Western historians' use of the Indian village in political analogy.

craftsmanship in the writing, but the larger design, the model eventually constructed, does not quite satisfy.

We are often left unsure about the explanations Spate offers to phenomena that seem to be of concern to him, chiefly areal variations in man's use of the land. Take the way he organizes a regional section of a regional chapter. There is deft sketching, more descriptive than explanatory, of geographical outlines, and some consideration of reasons, not excluding historical ones, for treating as a region the area described. Then the physical ensemble--land surface and climate, and, less prominently, and partially correlated with them, vegetation and soils--is presented as implicit cause of such interrelated effects as extent and mix of cultivation, and population density. Certain not easily defensible assumptions about the uniformity and constancy of Indian crop repertoires and technical capabilities, not to mention the absence of friction of distance to market, appear to underlie these explanations, but they are not spelled out. One is not told about the spatial limits of these predictive generalizations, nor if they do not hold everywhere, why this should be so, nor how the spatial limits may or may not be related to the proffered regionalization. More than once, too, Spate appears to be attempting explanatory jumps that exceed the reach of the "neo-determinist" philosophy he has espoused elsewhere.[21]

In all of this, Spate is reflecting accurately the contemporary milieu of Indian geography, although by judiciously screening and selecting source material from a very wide range, he has transmuted it into a masterly descriptive geography. One can gauge the extent of Spate's achievement by scanning Chatterjee's two volumes on the progress of geography in India,[22] which conveniently list virtually everything published in the organs of Indian academic geography

---

[21]"Extreme population pressure and perhaps the rival traditions of utmost Hindu orthodoxy and of dissent, have given the new state [Kerala] an unusually extremist political cast" (Spate and Learmonth, 1967), p. 673); the very marked contrast in Andhra between the population densities of Telengana and Rayalaseema on the one hand and the coastal districts on the other "seems to indicate a possible source of weakness in the new state" (ibid., p. 716). Or conflicting explanations are proposed: "Despite, or rather because of, the fertility and good rain of most of the region [Upper Gangetic Plain], population presses very hardly on resources" (ibid., p. 555). An exposition of Spate's philosophy is provided in O. H. K. Spate, "The Compass of Geography, Inaugural Lecture Delivered at the Australian National University, 8 September 1953," in O. H. K. Spate, Let Me Enjoy: Essays, Partly Geographical (London: Methuen, 1966), pp. 91-120.

[22]S. P. Chatterjee, Fifty Years of Science in India: Progress of Geography (Calcutta: Indian Science Congress Association, 1963); Progress of Geography in India (1964-1968) (Calcutta: Indian Science Congress Association, 1968).

since the 1920's. The pervading impression of banality that a reading of these volumes conveys cannot be entirely blamed on the editor's attempt to summarize in a line or two every article and book he cites. But we may question whether Spate's descriptions of landscape, the uncertain ecological models, the gazetteer-like locating of industrial activities have much to say to the Indianist.

In any case, there are few connecting links that would allow the student of Indian society to make use of them. The relationships implied are not investigated by Spate within a framework of acquaintance with Indian social and cultural-ecological processes (a contrast to the discussion of terrain features, which always brings in consideration of geomorphic processes). Here the fault does not lie entirely with a lack of source material. Spate praises both Geddes and Mukerjee, but their suggestive comments about possible links between social organization and such features as population distribution and settlement forms appear to have been deliberately screened out.

A curiously anachronistic weakness in the revised edition is the uneven treatment of spatial relationships. Spate's management of these is extraordinarily skillful in the chapter on historical geography, but his well-known concept of the "perennial nuclear region" and his expert handling of historical circulation flows are not reflected in cultural and social analysis at other scales and in other periods. Thus, he will have little to do with the notion of contemporary city-centered regions.[23] His "empirical" regional schema, retained from the first edition, is in the end based very largely on land surface structure, from the familiar tripartition of first order regions (mountain, plain, plateau) all the way down to sixth order "stows" and "features," sometimes no more than a few hundred square miles in area. Since what follows his one brief explication of the schema is a region by region <u>human</u> geography, it is hardly surprising that an anthropologist resorting to it unawares might be confused by it.[24] Indianists who are not geographers may then be pardoned for not wanting to use regional frameworks that they find unrecognizable and that appear to bear little relation to the patterns that interest them.

The most recent South Asian regional geography of broad scope is the massive volume conceived and edited by R. L. Singh of Banaras Hindu University.[25] Like Spate's book, it reflects the work and interests of a large body of

---

[23]Spate and Learmonth (1967), pp. 407 ff.

[24]Mandelbaum (1970), p. 391.

[25]R. L. Singh (1971).

Indian geographers, biased toward the particular concerns of Professor Singh and his Varanasi school. But it does not offer a similar integration, displaying, on the contrary, a serious structural defect.

Having passed in quick review the gist of twentieth-century geographic thought concerning "the region," Singh proposes his own regionalization of India. He recognizes the need to leave such schemata pragmatic in intent, convenient simply as a framework of study, necessarily imprecise as to boundaries, and subject to a good deal of reasonable disagreement. He does give us some odd pieces, over which one might quarrel, but the scheme is on balance neither better nor worse than many of its kind. The regions have been delineated by considering both (physical) geographical factors and "the historico-cultural organization of space," and the inadequacies of the larger administrative units for this purpose are recognized--they can "carry us astray from the geographic realities."[26] The scheme thus avoids the peculiarities of the one followed by Chatterjee[27] in which the boundaries of physiographic regions are seen to coincide providentially with present state boundaries in an unusually large number of cases.

Having completed this admittedly improvisatory exercise, Singh dismays us by snapping the frame over the map of India and locking it tightly in place. The "regions" thus created are treated as separate, almost completely closed territories. One would like to say "systems" but the possible systemic relations among the "factors" described are barely examined, and few interrelationships are implied: the factors are presented "merely to give a balanced regional account."[28] The interaction and interdependence between, say, central and eastern Uttar Pradesh are obliterated, and the form and functions of the Delhi metropolitan area are treated in a page or two in a chapter on the Punjab Plain Region.

In thus clamping without apparent purpose on arbitrary frame of regions over the area to be studied, Singh is by no means alone among Indian geographers. Two recent geographies provide further examples of this practice, E. Admad's study of Bihar,[29] where the principles followed in devising the regions remain

---

[26] Ibid., p. 33.

[27] S. P. Chatterjee, "Physiography," in The Gazetteer of India, Vol. I: Country and People (New Delhi: Publications Division, Ministry of Information and Broadcasting, 1965).

[28] R. L. Singh (1971), p. 28.

[29] Enayat Ahmad, Bihar, A Physical, Economic and Regional Geography (Randhi: Ranchi University, 1965).

obscure, and Deshpande's able work on Maharashtra,[30] addressed to a lay audience. Deshpande shows an interest in social circulation and even names the traditional Maharashtrian sub-regions. But he abandons them immediately, not asking how they came into being or what relevance they might have today, and replaces them with the "physical-economic" regions of a now stale convention.

Singh's book does have refreshing qualities. The interest of the Varanasi school in rural settlement and urban morphology is expressed in a consistent documentation with maps of the arrangement of villages on the land, of crops on village lands, of caste groups and institutional structures in village residential areas. Systematic relations of the kind suggested by Mukerjee may, however, be left unexamined, although there are notable exceptions, such as the reconstruction of the historical pattern of caste-differentiated settlement in northeastern Uttar Pradesh, which owes much to the work of K. N. Singh. Some of the more perfunctory treatments of the topic may have been a disappointment to the editor, whose other work suggests an appreciation of land and place, and a feeling of being at home with his area. These are encouraging signs, as are indications of interest in the popular history of Indian localities. Offsetting these is an evident reluctance to discuss current and potential sources of social and cultural tension; the deference to state sensitivities in such matters appears to be complete.

The reluctance in the geographic convention dominant in India to overcome distance between observer and observed--people and their environment--has provided an unusually receptive niche for the methodological developments of the past two decades, popularly subsumed under the rubric of "the quantitative revolution." The indistinctness of social and economic processes in the descriptive models of the geographer could be modified by the successful search for such processes in the spatial configuration of phenomena already observed by the agencies collecting intelligence for the state. The unusually rich ore of the censuses and other compendia of statistics was ready to be smelted by the new machines of the computer technology. There is, to be sure, a long time lag in the diffusion of these innovations to geographical practice in India, not least because of their expense, and one is aware of only modest beginnings there as yet. Nonetheless, there has been a rapid rise in statistics-based geography, using cruder and slower methods than those permitted by the computer, and the omens

---

[30] C. D. Deshpande, Geography of Maharashtra (New Delhi: National Book Trust, 1971).

for an imminent recapitulation of the "transatlantic" experience in geography are not to be mistaken.

The most vigorous statement of the programmatic possibilities thus opened for South Asian geography has come from the group that is most uncompromising in its dissociation from any sort of dweller's or stranger's geography, the school of Brian Berry and his students at Chicago. One must be wholeheartedly grateful to him for trying to shatter the hard crust of "uniform regions" that has so immobilized thinking in Indian geography. The complex spatial patterns of interconnection and flow at a variety of scales that some earlier geographers had discerned are now expressed, we are told, in the stark, almost brutally simplified outlines yielded by Berry's geographical analysis. Although Berry himself has appeared at times to be qualifying the validity of these exercises, certainly impressive as demonstrations of mechanical muscle flexing, inordinate value is still claimed for them. Recognizing that methodological exploration and reassessment are continuing, I want to examine in some detail a few representative works by Berry and his disciples for what they have to say to the Indianist. It should be understood that this cannot do justice to the full range of themes pursued by this group and, more importantly, perhaps, that they do not represent themselves as having something special to say to this particular audience, any more than, say, Reclus did.

It is precisely the judgment, to which it has been least exposed, of its contribution to an understanding of South Asia that finds this body of work at its most vulnerable. Typically, in these analyses, observations are transformed and manipulated in extraordinarily complicated ways. There is then considerable question as to whether the mathematical results can properly be translated back into a simple description of relationships among the original terms. Yet time after time, this is what we are offered with very little apology; it is these alleged re-translations that are transmitted as substantive findings to the lay world of South Asian scholarship. But these interpretations, far from being logical retranslations, are mostly plausible intuitions. Being couched in western sociological and economic terms, they often turn out to be simply wrong, to judge by the available empirical evidence. This evidence is not always abundant, but as a rule Berry and his students have declined to scrutinize it with care.

This is not the occasion--nor do I have the competence--for a comprehensive review of this unusually interesting corpus of work. The point I wish to make is that it is the particular location of Indian geography in intellectual space --remote from a major part of its subject, man in India, and also remote from those disciplines that do focus on that subject--that permits, even fosters, the kind of culturally-biased distortion that can so easily ensue from factor analytic modes of geography. I shall try to demonstrate this by scrutinizing a few examples.

Berry's own most substantial work on India, is his monograph on commodity flows.[31] This is an attempt to describe the grand spatial structure of the Indian economy. One can only note in passing a number of methodological eyebrow-raisers, about which the author remains silent. What is the effect of using "trade blocks," as sources and destinations of commodities, which vary enormously in area and population? Since each of the four largest metropolises is separately treated in the commodity flow statistics, while other large centers are merged in their state hinterlands (e.g., Kanpur within Uttar Pradesh), how can one properly differentiate the role of the two kinds of cities? How is "the spatial structure of the economy" affected by factoring as equivalents the flows of commodities whose total quantity and value vary widely? How can averaging district factor scores by trade blocks be defended when the districts contribute in very different proportions to the properties of the block? One is persuaded that however importantly these questions must bear on interpretation of the results, they have no real interest for the author.

To judge from evidence available in the work, conclusions are reached quite subjectively, although they may be intuitively sound. Clearly, the method is all. Whatever the theoretical interest of canonical correlation of structural and commodity flow factors, a student of India would require extraordinary faith to accept Berry's translation of his final matrix.

The difficulties that surround interpretation of the massive factor analysis that is Berry's trademark are illustrated by the work of his student, Qazi Ahmad.[32] Ahmad's aim is to determine the "critical properties" of Indian cities as described by 62 variables covering "a variety of the relevant aspects of urban structure in India."[33] However, the relevance of the variables, except for a handful of locational ones, is not discussed, their greatly varying relia-

---

[31] Brian J. L. Berry, Essays on Commodity Flows and the Spatial Structure of the Indian Economy, University of Chicago, Department of Geography Research Paper No. 111 (1966).

[32] Qazi Ahmad, Indian Cities: Characteristics and Correlates, University of Chicago, Department of Geography Research Paper No. 102 (1965).

[33] *Ibid.*, p. 18.

bility is not considered, and some critical variables that might have been suggested by descriptions of Indian cities are omitted.

After submitting his matrix of variables for 102 cities to principal components analysis and multiple discriminant analysis, Ahmad is able to show how the cities group on the basis of similarity in respect to statistically separated ("independent") dimensions of variation of the variables used. He seems puzzled by the fact that cities and town groups exhibit marked spatial clustering, finally combining in a northern and southern block, but does not ask himself whether these clusters might be fictitious, an effect produced simply by retaining spatial variables in the matrix. Nor does he suggest that there may be in India "regional" socio-economic structures, palely reflected in these city groupings. In one striking instance, he is reluctant to accept a "cultural" explanation for the strongest element of urban differentiation, the north-south variation in sex ratio (i.e., there are fewer females per male in northern cities). He prefers to emphasize selective migration and the extremely small proportion of Muslims in the south as compared with the north, noting that "the Muslim population, in general, suffers from a highly unbalanced sex ratio (more males to females)."[34] There should then be more females than males in most northern rural areas, and a marked difference, district by district, between Muslim and other sex ratios. But there are not, and one sees how, in South Asian geography, the dweller is often made stranger by the conventions of the prevailing paradigm.

This may appear to be flailing at a straw man, but it is surely in order if such work represents a direction in which South Asian geography is likely to go.[35] The evident lack of concern to adapt Western categories of urban analysis to the context of Indian civilization augurs poorly for wise planning applications or a sound reformation of Indian geography. And yet, are not these problems of cultural translation acknowledged, accepted, indeed confronted by Berry and his fellow practitioners? One is almost persuaded of this in reading some apologetic passages in Berry's work on factorial ecology of Indian cities,[36] and in

---

[34] Ibid., p. 118.

[35] One is not at all reassured by the recent appearance of a particularly mindless piece of "gigoism," judging by the variables chosen and the interpretation of the results, namely, Frank E. Horton, Harold McConnell, and Ranjit Tirtha, "Spatial Patterns of Socio-Economic Structure in India," Tijdschrift voor Economische en Sociale Geografie, Vol. 61 (1970), pp. 101-113.

[36] Brian J. L. Berry and Philip H. Rees, "The Factorial Ecology of Calcutta," The American Journal of Sociology, Vol. 74 (1969), pp. 445-491; Brian

noting welcome evidence (in the Berry and Spodek article) that the authors have checked some of their conclusions in the real world of the cities involved.

There is still some squeezing and wrenching of statistical results to make them fit an a priori model. A student of India might be particularly reluctant to accept the repeated assertion, based on very little evidence, that Calcutta in 1961 is temporally transitional, not just typologically, between the pre-industrial and industrial. Nor would he feel confident that in the polyethnic metropolitan situation of Calcutta, the authors can correctly identify the dimension representing "status" because that is what it appears to be in Chicago. Berry's assumptions as to what "industrial" cities must be like require that the socioeconomic and ethnic bases of differentiation be separate. Since they do not appear to be in Calcutta, then the city _must_ be transitional and the separation will probably occur. Thus, too, the caste basis of social area differentiation must be declining in large Indian cities, despite conflicting evidence.

Is the comparative factor analysis of cities a useful procedure then? In a collection of work on the theme edited by Berry, Susan Landay expresses strong reservations because of the distortions of culture-bound categories that the procedure requires and the absence of suitable theory dealing with cities in different cultures.[37] She particularly notes the fallacy of equating modernization in non-Western cultures with "westernization." Conceding a certain justice to such criticisms of comparative factorial ecology and to the charge of Western ethnocentrism, Berry is satisfied to dispose of them in these words: "To these charges we can only answer that the phenomenon of cultural diffusion today leads one to doubt that the thesis of cultural integrity and _deliberately_ obfuscating ethnocentrism have much force, particularly in urban studies."[38] So much for foreign cultures and their study: all the earth _is_ to be Pasadena!

In the essays just described we will search in vain for some invitation to go into the field in order to look carefully at the cities involved, to examine the

---

J. L. Berry and Howard Spodek, "Comparative Ecology of Large Indian Cities," in Brian J. L. Berry (ed.), "Comparative Factorial Ecology," _Economic Geography_, Vol. 47 (1971), Supplement, pp. 266-285.

[37]Susan Landay, "The Ecology of Islamic Cities: The Case for the Ethnocity," in Brian J. L. Berry (ed.), "Comparative Factorial Ecology," _Economic Geography_, Vol. 47 (1971), Supplement, pp. 303-313.

[38]Brian J. L. Berry, "Introduction: The Logic and Limitations of Comparative Factorial Ecology," in Brian J. L. Berry (ed.), "Comparative Factorial Ecology," _Economic Geography_, Vol. 47 (1971), Supplement, pp. 209-219. Citation on p. 219.

nature of the social being that is described in the census categories used as variables, or to consider the possibilities of error arising from the way in which the data in these categories have been ascertained. The thread of experience linking the geographer with people in the cities and on the land becomes ever more tenuous.

Censuses and other statistical sources can of course be used in rewarding ways in South Asian studies. Schwartzberg's work on castes and a monograph in which he assesses the reliability of census occupational statistics show the care with which one can and should handle data of this kind.[39] His work also stands out as a distinguished exception to the remoteness from the subject that has been so pervasive a feature of Indian geography as done by Westerners, and, indeed, by many Indians. By bringing into geography what South Asian specialists in other disciplines had been constructing in the way of theory about Indian society and culture, Schwartzberg has been able to offer an important reciprocation, suggesting modifications, refinements, and elaborations in which are models of a complex Indian civilization rather than patched over structures cast in an ethnocentric Western mold.

Schwartzberg is not the only exception. Some middle-ranked Indian geographers have moved in directions foreshadowed by Radhakamal Mukerjee in an attempt to tie together ecology, settlement, and social structure, through time, for particular Indian landscapes.[40] One can have high hopes for some of the younger South Asian geographers, recently or still studying abroad, who are catching on to the importance of looking afresh as dwellers at their own land. In the past decade, too, a handful of European geographers have made considerable

---

[39] Joseph E. Schwartzberg, "The Distribution of Selected Castes in the North Indian Plain," Geographical Review, Vol. 55 (1965), pp. 477-495; "Caste Regions of the North Indian Plain," in Milton Singer and Bernard S. Cohn (eds.), Structure and Change in Indian Society (Chicago: Aldine, 1968), pp. 81-114; Occupational Structure and Levels of Economic Development in India: A Regional Analysis, Census of India 1961, Monograph No. 4 (New Delhi: Manager of Publications, 1969).

[40] E.g., A. B. Mukerji, "Succession of Cultural Landscapes in Telengana Reddi Villages," Indian Geographical Journal, Vol. 39 (1964), pp. 42-58; "Cultural Landscape of a Telengana Village: A Morphogenetic Study," Geografia (Karachi), Vol. 3 (1964), pp. 11-42; "Musulman House Type in the Meerut Area," Folklore, Vol. 8 (1967), pp. 116-130; Kashi Nath Singh, "Evolution of Early Medieval Politico-Cultural Regions and Urban Foci in Eastern Uttar Pradesh, India," Geographical Outlook, Vol. 4 (1965), pp. 23-40; "The Territorial Basis of Towns and Village Settlement in Eastern Uttar Pradesh, India," Annals of the Association of American Geographers, Vol. 58 (1968), pp. 203-220.

strides in trying to understand particular villages or pays as complex social, ecological, and spatial systems.[41]

Other Directions

In what other directions might geography move in order to establish a secure place within South Asian area studies? Several large related themes that remain central to the geographic tradition deal with questions which students in other disciplines are continually seeking to answer. Geographers working in South Asia have hitherto either neglected these questions or given them only scant attention. Sometimes it is scholars in other disciplines who have done the critically important work leading to a fresh consideration of the old questions. Of course, even a geographer may regard such a development with equanimity; the terms of this discussion rather than personal preference call for me to discriminate in regard to disciplinary territoriality.

The effect of environment on man is one large, ancient question of this sort, and however it ought to be rephrased to discriminate among physiological effects and cultural perceptions, it is one that geographers have shied away from too long. We find then that intellectual gadfly, Nirad Chaudhuri, going unchallenged by geographers in his assertion that India's tropical environment has been a cause of degeneration of vigorous immigrant cultures,[42] while it is an anthropologist who appears as the author of a careful study of cultural institutions and practices that seem intended to alleviate heat stress in northern India.[43] Perhaps these diverse reminders of the interest that the question holds will encourage geographers to attempt its systematic investigation, paying attention to folk perceptions, customs, calendars, rhythms--the coping mechanisms, especially in domestic architecture, that are often pushed aside for the

---

[41] E.g., Jacques Dupuis, Madras et le nord du Coromandel: étude des conditions de la vie indienne dans un cadre géographique (Paris: Adrien-Maisonneuve, 1960); Ulf Hoppe, Chintakunta: die agrargeographische Struktur eines Dekkan-Dorfes, Arbeiten aus dem Geographischen Institut (Frankfurt am Main, 1966); Dirk Bronger, "Der sozialgeographische Einfluss des Kastenwesens auf Siedlung und Agrarstruktur im Südlichen Indien," Erdkunde, Vol. 24 (1970), pp. 194-207.

[42] Nirad C. Chaudhuri, The Continent of Circe (New York: Oxford University Press, 1966).

[43] Jack M. Planalp, Heat Stress and Culture in North India, U.S. Army Research Institute of Environmental Medicine, Special Technical Report (Natick, Mass., 1971).

adoption of inappropriate foreign technologies. One might note how both the social calendar and the testimony of landscape painting, sculpture, and poetry express the sustained exuberance of the northern fair weather season and celebrate the vitality released by the onset of the rains. These remarks are not to dismiss much that has been done in Indian geography under the rubric of geomorphology and climatology. It is the possible connections between this work and the geographic study of the civilization of India that have scarcely been pursued and so the work remains largely irrelevant for the field of Indian studies. Some recent collations of diverse physical environmental and socio-economic data have potential of this kind and ought to be better known outside the geographer's field: the English edition of the National Atlas of India, now appearing in separate sheets, and the admirable work of the French Institute at Pondicherry, whose series of maps and monographs on vegetation and land use for much of India provide a valuable source for Indian studies at a variety of scales.

Perhaps geography has otherwise had little to say to the South Asian area specialist because it has often taken as its model relationships between religious belief, societal values, and ecological institutions that have been losing acceptance elsewhere. The controversy launched by the anthropologist Marvin Harris over the possible relation of the "sanctity" of cattle in Hinduism to Indian agricultural ecologies[44] has caught geographers, including most Indian ones, flat-footed, their prejudices showing. These same prejudices, a concomitant of orientation to both Western technology and state purposes, have resulted in the failure of geography to present a sound historical balance-sheet for Indian agricultural ecosystems. Such an accounting would illuminate the remarkable solutions of Indian folk technology to the problems of sustaining an intensive agriculture in very difficult physical environments; compare, for example, the population densities and the complexity of social and economic organization of the Deccan plateau, past and present, with what is found in climatically analogous regions anywhere else.

In this critically important historical review of ecological systems, basing the analysis of the past relationships of man to land on the evidence inscribed in the landscape, South Asian geography has remained deficient. Some fine work has been done toward reconstructing the history of hydrological systems, especially the lower flood plains of the two major rivers, with the Indus receiving

---

[44] Marvin Harris, "The Cultural Ecology of India's Sacred Cattle," Current Anthropology, Vol. 7 (1966), pp. 51-66.

most attention recently.[45] Much remains hardly touched, though dimly discernible; pressure of man on vegetation, extinction, but also preservation, of fauna, an ancient organizational shift of cultural cores from thin-soiled uplands to flood-plain margins, the filling of the land by internal colonizing movements that are recorded in folk history (especially caste histories), but have yet to be elaborated by careful landscape analysis.

In recent years, research of this kind has suffered from unnecessary handicaps in most of the region (the peripheral lands--Ceylon, Nepal, Afghanistan--are exceptions). The restricted circulation of large-scale maps and the general unavailability of aerial photography have been a serious problem. There is no dearth of conspicuous question marks in the field--the large living mound-villages of Western Uttar Pradesh; the frequent, more widespread occurrence of ruin heaps (kheḍā) in the level plain; the rectilinear, cardinally oriented system of village embankments (ahars) in the core of ancient Magadha; the distribution of nucleated and hamletted settlement in eastern and central Uttar Pradesh, to name only a few North Indian features of interest. It is frustrating to be unable to explore, in the systematic way that the available technology should permit, the questions these features raise.

Where geographers, not working alone but together with other South Asian specialists, can do most fruitful work (some examples of which have already appeared) is in elucidating the full, rich complexity of the spatial structures of Indian civilization. Factor analytical modes will not do it. Questions of growth and decay within a long-lived urban system, or of regional differentiation in language and social institutions require models of the "cellular" spatial structure of society, of the pattern of interaction among village-based kin, of their perceptions of place and social network, and of important regional variations in such patterns. We may note that Skinner's exposition of the relation between social networks in rural China and market networks[46] has been taken as a general model of such relationships. Yet northern India constitutes a striking exception: there the town or marketing central place often represents a void

---

[45] H. T. Lambrick, "The Indus Flood-Plain and the Indus Civilization," Geographical Journal, Vol. 133 (1967), pp. 483-495; Herbert Wilhelmy, "The Shifting River: Studies in the History of the Indus Valley," Universitas, Vol. 10 (1967), pp. 53-68; "Indusdelta und Rann of Kutch," Erdkunde, Vol. 22 (1968), pp. 177-191; "Verschollene Städte im Indus Delta," Geographische Zeitschrift, Vol. 56 (1968), pp. 256-294.

[46] G. William Skinner, "Marketing and Social Structure in Rural China: Part I," Journal of Asian Studies, Vol. 24 (1964), pp. 3-44.

in the lattice of village marriage linkages. One looks then for convergent contributions from geographer and anthropologist, as exemplified in work on the relationship between social-political organization and settlement pattern in eastern Uttar Pradesh.[47]

Many themes that invite investigation through the analysis of circulation flows in India have fallen into the interstices separating the social science disciplines, as it were, and lie ready to be picked up and developed by the geographer: the spatial dynamics of ideas, contemporary political and religious movements, the diffusion of culture by the mass media; bilingualism and lingue franche; changing morphology of language transition zones and potential language barriers to interregional migration and communication; access to the political system; cities as cultural reintegrators. Thus back to the questions of interaction field, sense of home, cues to identity. How do the people of pays view themselves and others? How do those of larger regions? Research might explore folk literature and regional literature, perhaps developing a whole matrix of reciprocal stereotypes, and thence move to a full political geography of South Asia, which hardly exists at present, even in rudimentary form. One can make a beginning in social geography, as I have tried to do, by using census data to develop statistical surfaces of socio-cultural significance, a form of exploration that simply tries to ask discriminating questions regarding extent and intensity of phenomena that have been investigated at a few locations by scholars from other disciplines.[48]

Summary

How might one then summarize the "response" of geography to the "challenge" of India? My inclination is to score it thus far as a qualified failure. The

---

[47] Kashi Nath Singh, "Evolution of Early Medieval Political-Cultural Regions and Urban Foci in Eastern Uttar Pradesh, India," Geographical Outlook, Vol. 4 (1965), pp. 23-40; "The Territorial Basis of Towns and Village Settlement in Eastern Uttar Pradesh, India," Annals of the Association of American Geographers, Vol. 58 (1968), pp. 203-220; Richard G. Fox, "Rajput 'Clans' and Rurban Settlements in Northern India," in Richard G. Fox (ed.), Urban India: Society, Space and Image (Duke University Program in Comparative Studies on Southern Asia, Monograph No. 10, 1970), pp. 167-185; Kin, Clan, Raja and Rule: State-Hinterland Relations in Pre-Industrial India (Berkeley: University of California Press, 1971).

[48] One looks forward with great interest to the forthcoming publication of a social and historical atlas of South Asia, compiled under the energetic direction of Joseph Schwartzberg, that will present some of the distributional correspondences of social and ecological patterns that now excite the curiosity of the connoisseur.

field has been undermanned, but its strategy, the questions it has chosen for its focus and the methods used have been instrumental in bringing this result about. This is not meant to be an assessment of achievement "in the service of the state," which is for many Indian geographers today a necessary justification and sufficient reward for what they do. If one were asked to offer a prescription, it would be that South Asian geography should become more "South Asian," more conscious of its own categories of behavior and thought, of its peoples and their worlds, than it can be by seeing South Asia with Western eyes. For this purpose, South Asian geographers must learn to work closely with other students of South Asian societies and civilization, and they must learn to work closely with people on the land and in the cities. A useful measure of progress in this respect might be the degree to which cross-cultural studies within the South Asian region will come to be a matter of course, carried out by multilingual geographers working back and forth between their own and other linguistic-cultural regions. The chief task of the foreign and Indian "foreign area specialist" then might be to mediate the flow of ideas emanating from such studies to and from the Western and other culture regions represented at this symposium. That, too, is a prospect that we may well regard with equanimity.

CHAPTER 7

GEOGRAPHICAL PERSPECTIVES ON THE MIDDLE EAST: THE PASSING OF THE ECOLOGICAL TRILOGY[1]

Paul Ward English
University of Texas

Middle Eastern society until recently was viewed as an ecological trilogy, a system of settlement composed of three types of community--the city, the village, and the tribe. Each of these components formed a distinctive human environment; each was rooted in a different physical environment. All shared in the moral order of Islam.[2] While recognizing the difficulties inherent in

---

[1]The term "Middle East" is implicitly enthnocentric, but common usage probably makes it preferable to alternatives like the Near East, the Nearer East, Hither Asia, Southwest Asia, Swasia, or Southwest Asia and North Africa. See Marshall G. S. Hodgson, "The Interrelations of Societies in History," Comparative Studies in Society and History, Vol. 5 (1963), pp. 227-250. The region here discussed includes all of desert North Africa south to approximately 15° latitude, the Arabian Peninsula, the Levant, and the high plateaus of Turkey and Iran as far eastward as Afghanistan. The Middle East covers more than 7 million square miles, approximately 90 per cent of which is uncultivated, and is occupied by some 200 million people, 95 per cent of whom are Muslim. A summary of the literature debating the perimeters of the region is provided in Roderic H. Davison, "Where Is the Middle East?" in Richard H. Nolte (ed.), The Modern Middle East (New York: Atherton, 1963), pp. 13-29.

[2]Alternate approaches define the Middle East as the "dry world," the "world of Islam," the "Arab world," or a "mosaic of peoples." Discussions of the merits of these definitions and the difficulties of setting the bounds of the region preface nearly every general text. For example: Gabriel Baer, Population and Society in the Arab East (London: Routledge and Kegan Paul, 1964); Morroe Berger, The Arab World Today (Garden City, N.Y.: Doubleday, 1962); William C. Brice, Southwest Asia: A Systematic Regional Geography (London: University of London Press, 1966); Carleton Coon, Caravan: The Story of the Middle East (New York: Holt, Rinehart and Winston, 1958); George B. Cressey, Crossroads: Land and Life in Southwest Asia (Philadelphia: J. B. Lippincott, 1960); W. B. Fisher, The Middle East: A Physical, Social and Regional Geography (London: Methuen, 1971); Stephen H. Longrigg, The Middle East: A Social Geography (Chicago: Aldine, 1963); Raphael Patai, Golden River to Golden Road (Philadelphia: University of Pennsylvania Press, 1967); Xavier de Planhol, The

generalizations based on a few case studies loosely scattered over a vast area, many scholars implicitly or explicitly accepted this tripartite division of Middle Eastern society as reflecting three "ways of life" sufficiently distinctive to merit typological consideration.[3] Thus a body of literature emerged debating and discussing the essential characteristics of the Islamic city, the Middle Eastern village, and the tribal encampment as they have derived from centuries of ecological experimentation, from the culture and institutions of Islam, and from the differing social and economic realities bounding the horizons of these communities at different times.

One value of this paradigm lay in its capacity to generate comparative discussions among the bewildering variety of scholars who have contributed to Middle Eastern studies. Urbanists focused on the role of the mosque and its religious intelligentsia ('ulamā'), the structure and administration of the bazaar, the system of quarters (ḥārah) and streets (darb), and more specifically on the role of minorities, brotherhoods, and guilds in cities as diverse as Cairo, Baghdad, Marrakesh, and Shiraz.[4] Although many scholars continued to stress

---

World of Islam (Ithaca: Cornell University Press, 1959); idem, Les fondements géographiques de l'histoire de l'Islam (Paris: Flammarion, 1968); C. A. O. van Nieuwenhuijze, Sociology of the Middle East (Leiden: E. J. Brill, 1971).

[3]Among them: Elizabeth E. Bacon, "A Preliminary Attempt To Determine the Culture Areas of Asia," Southwestern Journal of Anthropology, Vol. 2 (1946), pp. 117-132; Xavier de Planhol, "Regional Diversification and Social Structure in North Africa and the Islamic Middle East," in Richard Antoun and Iliya Harik (eds.), Rural Politics and Social Change in the Middle East (Bloomington: Indiana University Press, 1972), pp. 103-117; Paul Ward English, "Urbanites, Peasants, and Nomads: The Middle Eastern Ecological Trilogy," Journal of Geography, Vol. 66 (1967), pp. 54-59; Raphael Patai, "Nomadism: Middle Eastern and Central Asian," Southwestern Journal of Anthropology, Vol. 7 (1951), pp. 401-414; idem, "The Middle East as a Culture Area," Middle East Journal, Vol. 6 (1952), pp. 1-21; Louise Sweet, Tell Toqaan: A Syrian Village (Ann Arbor: University of Michigan Press, 1960), pp. 1-6, 228-230; Jacques Weulersse, Paysans de Syrie et du Proche-Orient (Paris: Gallimard, 1946), pp. 60-66. Among the texts noted above see especially Baer (1964), pp. 119-203; Coon (1953), pp. 171-259; and Patai (1967), pp. 39-83.

[4]General discussions of the Islamic City can be found in: Francisco Benet, "The Ideology of Islamic Urbanization," International Journal of Comparative Sociology, Vol. 4 (1963), pp. 211-226; Claude Cahen, "Mouvements populaires et autonomisme urbain dans l'Asie musulmane de moyen age," Arabica, Vol. 5 (1958), pp. 225-250, Vol. 6 (1959), pp. 25-46, 223-265; idem, "Zur Geschichte des Städtischen Gessellschaft im Islamischen Orient des Mittelalters," Speculum, Vol. 9 (1958), pp. 59-76; G. E. von Grunebaum, "The Structure of the Muslim Town," in his Islam and the Growth of a Cultural Tradition, 2nd ed. (London: Routledge and Kegan Paul, 1961), pp. 141-158; A. H. Hourani and S.

the uniqueness of each specific Middle Eastern city, generalized models of the morphology, value systems, and social organization of the traditional "Islamic city" were drawn, and simple classifications of "spontaneous" versus "created," "orthogenetic" versus "heterogenetic," and "sacred" versus "secular" cities were suggested.[5]

M. Stern (eds.), The Islamic City (Philadelphia: University of Pennsylvania Press, 1970); Susan Landay, "The Ecology of Islamic Cities: The Case for the Ethnocity," Economic Geography, Vol. 47 (1971), pp. 303-313; Ira M. Lapidus, Muslim Cities in the Later Middle Ages (Cambridge: Harvard University Press, 1967); idem (ed.), Middle Eastern Cities: A Symposium on Ancient, Islamic, and Contemporary Middle Eastern Urbanism (Los Angeles and Berkeley: University of California Press, 1969); M. Lombard, "L'Évolution urbaine pendant le haut moyen âge," Annales: économies, sociétés, civilisation, Vol. 12 (1957), pp. 7-28; G. Marçais, "La conception des villes dans l'Islam," Revue d'Alger, Vol. 2 (1945), pp. 517-533; Xavier de Planhol (1959), pp. 1-41; J. Weulersse, "La primauté des cités dans l'économie syrienne," Comptes rendus du congrès international de géographie, Leiden 1938, Vol. 2 (1938), pp. 233-239.

Case studies of Middle Eastern cities include: Janet Abu-Lughod, Cairo: 1001 Years of the City Victorious (Princeton: Princeton University Press, 1971); Fredy Bémont, Les villes de l'Iran (Paris: Fabre, 1969); Morroe Berger (ed.), The New Metropolis in the Arab World (New Delhi: Allied, 1963); Jacques Berque, "Médinas, villeneuves et bidonvilles," Cahiers de Tunisie, Vol. 6 (1958), pp. 5-42; John I. Clarke, The Iranian City of Shiraz (Durham: Department of Geography, University of Durham, 1963); idem and B. D. Clark, Kermanshah: An Iranian Provincial City (Durham: Department of Geography Research Series No. 10, 1969); Marcel Clerget, Le Caire: Etude de géographie urbaine et d'histoire économique, 2 Vols. (Cairo: Imprimèrie E. and R. Schindler, 1934); Paul Ward English, "Culture Change and the Structure of a Persian City," in Carl Leiden (ed.), The Conflict of Tradition and Modernism in the Muslim Middle East (Austin: University of Texas Press, 1966), pp. 32-48; idem, "The Traditional City of Herat, Afghanistan," forthcoming in L. Carl Brown (ed.), Urban Planning and Urban Prospects in the Near East and North Africa (Princeton: Princeton University Press, 1973); John Gulick, "Baghdad: Portrait of a City in Physical and Cultural Change," Journal of the American Institute of Planners, Vol. 33 (1967), pp. 246-255; idem, Tripoli: A Modern Arab City (Cambridge: Harvard University Press, 1967); Georges Hamdan, "The Growth and Functional Structure of Khartoum," Geographical Review, Vol. 50 (1960), pp. 21-40; C. S. Hurgronje, Mekka in the Latter Part of the 19th Century (Leiden: E. J. Brill, 1931); Roger Le Tourneau, Fès avant le Protectorat (Casablanca: Institut des hautes études marocaines, 1949); idem, Les villes musulmanes de l'Afrique du Nord (Algiers: La Maison de livres, 1957); Laurence Lockhart, Persian Cities (London: Luzac, 1960); M. Pereira, Istanbul: Aspects of a City (London: Bles, 1968); K. Scharlau, "Moderne Umgestaltungen in Grundiss iranischer Stadt," Erdkunde, Vol. 15 (1961), pp. 180-191; Eugen Wirth, "Die Lehmhüttensiedlungen der Stadt Baghdad," Erdkunde, Vol. 8 (1954), pp. 309-316; idem, "Strukturwandlungen und Entwicklungstendenzen der orientalischen Stadt," Erdkunde, Vol. 22 (1968), pp. 101-128.

[5]G. E. von Grunebaum (1961), pp. 141-158; A. H. Hourani, "The Islamic City in the Light of Recent Research," in A. H. Hourani and S. M. Stern (1970), pp. 9-24; Ira M. Lapidus, "Muslim Cities and Islamic Societies," in Ira M.

Studies of Middle Eastern villages reflected a similar tension between accepted generalization and specific case study. Detailed ethnographies revealed complex variations in village social and economic organization and in the environmental basis of village life.[6] Yet the tendency was to view the village as a physically isolated, culturally homogeneous, subsistence settlement occupied by an inward-looking "peasant society." Corporate ownership patterns, kinship systems, a simple technology, and the rhythm of the agricultural cycle were seen as commonalities shared by the majority of Middle Eastern villages.

Tribal nomads were similarly compared in terms of their migratory regimes, their ecological adjustment to marginal resources, the composition and

---

Lapidus (1969), pp. 47-79; Edmond Pauty, "Villes spontanées et villes crées en Islam," Annales de l'institut d'études orientales, Algiers, Vol. 9 (1951), pp. 52-75; Planhol (1959), pp. 1-41; Robert Redfield and Milton B. Singer, "The Cultural Role of Cities," Economic Development and Cultural Change, Vol. 3 (1954), pp. 53-73.

[6]Studies of Middle Eastern villages include: Robert C. Alberts, Social Structure and Cultural Change in an Iranian Village (Unpublished Ph. D. dissertation, University of Wisconsin, 1963); H. Ammar, Growing Up in an Egyptian Village (London: Routledge and Kegan Paul, 1954); H. H. Ayrout, The Egyptian Peasant (Boston: Beacon, 1968); Harold Barclay, Buurri al-Lamaab: A Suburban Village in the Sudan (Ithaca: Cornell University Press, 1964); Brian Beeley, Rural Turkey: A Bibliographic Introduction (Ankara: Hacattepe University, 1969); Douglas C. Crary, "The Villager," in S. N. Fisher (ed.), Social Forces in the Middle East (Ithaca: Cornell University Press, 1955), pp. 43-59; Elizabeth W. Fernea, Guests of the Sheik (Garden City, N.Y.: Doubleday, 1965); Robert A. Fernea, Shaykh and Effendi: Changing Patterns of Authority in Southern Iraq (Cambridge: Harvard University Press, 1969); Anne H. Fuller, Buarij: Portrait of a Lebanese Muslim Village (Cambridge: Harvard Middle East Monograph Series No. 6, 1961); John Gulick, Social Structure and Culture Change in a Lebanese Village (New York: Wenner-Gren, Viking Fund Publications in Anthropology No. 21, 1955); John F. Kolars, Tradition, Season and Change in a Turkish Village, University of Chicago, Department of Geography Research Paper No. 82 (1963); idem, "Community Studies in Rural Turkey," Annals of the Association of American Geographers, Vol. 52 (1962), pp. 476-489; E. R. Leach, Social and Economic Organization of the Rowanduz Kurds (London: School of Economics Monographs in Social Anthropology No. 3, 1940); Abdulla Lutfiyya, Baytin: A Jordanian Village (The Hague: Mouton, 1966); Gavin Maxwell, People of the Reeds (New York: Harper, 1957); William Green Miller, "Hosseinabad: A Persian Village," Middle East Journal, Vol. 18 (1964), pp. 483-498; Joe E. Pierce, Life in a Turkish Village (New York: Holt, Rinehart and Winston, 1964); Paul Stirling, Turkish Village (London: Weidenfeld and Nicolson, 1965); Sweet (1960); Afif I. Tannous, "The Arab Village Community in the Middle East," Annual Report of the Smithsonian Institution for 1943 (1944), pp. 523-543; C. A. O. van Nieuwenjuijze, "The Near Eastern Village: A Profile," Middle East Journal, Vol. 16 (1962), pp. 295-308.

size of their herds, and the role of agriculture in their economy.[7] A continuum of nomadism was envisioned ranging from the "pure" nomadism of the camel-herding Bedouin through transhumance in the uplands of Anatolia and Iran, to small scale semi-nomadism on the periphery of cultivated areas. Ecological

---

[7]Studies of nomadic tribes include: Mohamed Awad, "The Assimilation of Nomads in Egypt," Geographical Review, Vol. 44 (1954), pp. 240-252; idem, "Nomadism in the Arab Lands of the Middle East," in UNESCO, The Problems of the Arid Zone (Paris: UNESCO, 1962), pp. 325-339; Fredrik Barth, Political Leadership among the Swat Pathans (London: School of Economics Monographs on Social Anthropology, No. 19, 1959); idem, Nomads of South Persia: The Basseri Tribe of the Khamseh Confederacy (London: George Allen and Unwin, 1961); idem, "Nomadism in the Mountain and Plateau Areas of South West Asia," in UNESCO (1962), pp. 341-355; Lloyd Cabot Briggs, Tribes of the Sahara (Cambridge: Harvard University Press, 1960); Robert Capot-Rey, "The Present State of Nomadism in the Sahara," in UNESCO (1962), pp. 301-309; Carleton Coon, "The Nomads," in S. N. Fisher (1955), pp. 23-42; Ian Cunnison, The Baggara Arabs (Oxford: Clarendon Press, 1966); E. E. Evans-Pritchard, The Sanusi of Cyrenaica (Oxford: Clarendon Press, 1963); C. G. Feilberg, La tente noire (Copenhagen: Nationalmuseets København, 1944); idem, Les Papis (Copenhagen: Nationalmuseets København, 1952); Klaus Ferdinand, "Les nomades afghans," in J. Humlum, La Géographie de l'Afghanistan (Copenhagen: Glydendal, 1959), pp. 277-290; C. Daryll Forde, "The Ruwala Badawin: Camel Breeders of Northern Arabia," in his Habitat, Economy and Society (New York: E. P. Dutton, 1963), pp. 308-327; William Ivons and Neville Dyson-Hudson (eds.), Perspectives on Nomadism (Leiden: E. J. Brill, 1972); Douglas L. Johnson, The Nature of Nomadism, University of Chicago, Department of Geography Research Paper No. 118 (1969), with an excellent bibliography; Rhoads Murphey, "The Decline of North Africa since the Roman Occupation: Climatic or Human?" Annals of the Association of American Geographers, Vol. 41 (1951), pp. 116-132; Alois Musil, Arabia Deserta: A Topographical Itinerary (New York: American Geographical Society, 1927); idem, The Manners and Customs of the Rwala Bedouins (New York: American Geographical Society, 1928); Johannes Nicolaisen, Ecology and Culture of the Pastoral Tuareg: With Particular Reference to the Tuareg of Ahaggar and Ayr (Copenhagen: National Museum, 1963); Robert N. Pehrson, The Social Organization of the Marri Baluch (New York: Wenner-Gren, 1966); Xavier de Planhol, "Geography, Politics and Nomadism in Anatolia," International Social Science Journal, Vol. 9 (1959), pp. 525-531; idem, "Les nomades, la steppe, et la forêt en Anatolie," Geographische Zeitschrift, Vol. 53 (1965), pp. 101-116; idem, "Aspects of Mountain Life in Anatolia and Iran," in S. R. Eyre and G. R. J. Jones (eds.), Geography as Human Ecology: Methodology by Example (London: Edward Arnold, 1966), pp. 291-308; Herbert F. Schurmann, The Mongols of Afghanistan (The Hague: Mouton, 1962); Thomas R. Stauffer, "The Economics of Nomadism in Iran," Middle East Journal, Vol. 19 (1965), pp. 284-302; Louise Sweet, "Camel Raiding among the Bedouins of North Arabia: A Mechanism of Ecological Adaption," American Anthropologist, Vol. 67 (1965), pp. 1132-1150; idem, "Camel Pastoralism in North Arabia and the Minimum Camping Unit," in Anthony Leeds and Andrew P. Vayda (eds.), Man, Culture and Animals: Animals in Human Ecological Adjustments (Washington, D.C.: American Association for the Advancement of Science, 1965), pp. 129-152; E. Wirth, "Das Problem der Nomadem im heutigen Orient," Geographische Rundschau, Vol. 21 (1969), pp. 41-51.

and cultural classifications engendered considerable discussion.

This treatment of city, village, and tribe as viable, identifiable sub-communities within the region naturally tended to call attention to interactions among them on the local, regional, and national levels. In truth, the exigencies of field research, limitations in historical documentation, the inadequacy of data bases, and weaknesses in research methodologies too often led to the analysis of single communities as isolates. Cities frequently were viewed as islands in a barren land, loosely linked by networks of roads and caravan trails to other urban centers. A clear dichotomy between "urban" and "rural" communities was often assumed, so that village ethnographies stress internal structure, not external relations. Tribes were ethnic and/or political groups who migrate periodically in prescribed patterns with only marginal contacts with sedentary communities. Despite these deficiencies in execution, the interactions between city, village, and tribe formed an important intellectual theme in the literature of the Middle East.[8]

These interactions among cities, villages, and nomadic tribes within the local region, a national system, and an international network delineated regional variations in Middle Eastern life. Networks of roads and caravan trails linked cities to one another, but the relationships between cities and villages were complex and varied. In some regions, notably Anatolia, a hierarchy of settlements organized the agrarian surplus of the hinterland to feed the population of cities like Istanbul. In Iran, the vast majority of villages were owned by the urban elite, who directly drained the hinterland of economic surpluses to sustain a graceful way of life. In southern Mesopotamia, Saudi Arabia, and parts of North Africa, villages paid tribute to or were owned by nomadic tribes. In these areas the distinction between peasant and nomad, between the desert and the sown, was often blurred by groups moving from one way of life to the other in the face of environmental uncertainty. In general, however, one point should be noted--life in the cities for a majority of its residents was not significantly better than life in the countryside. Indeed, urban death rates were so high that

---

[8]Geographers have played a major role in contributing to recognition of these problems. Paul Ward English, City and Village in Iran (Madison: University of Wisconsin Press, 1966); Kolars (1963); idem, "Types of Rural Development," in F. Shorter (ed.), Four Studies on Economic Development of Turkey (London: Frank Cass, 1967), pp. 63-87; Marvin W. Mikesell, Northern Morocco: A Cultural Geography (Berkeley: University of California Publications in Geography, No. 14, 1961). To the best of my knowledge no formal central place study has yet been published on the Middle East, although Michael Bonine's analysis of Yazd, Iran, will appear soon.

peasants and settled nomads often formed a significant percentage of the population. Common standards of living apparently spanned these three different living environments.

The integration of the modern Middle East into larger systems of trade, communication, politics, and economy has reshaped these interactions between city, village, and tribe.[9] Until the middle of the nineteenth century, these subgroups in the Middle East were relatively self-contained communities based on tribal affiliation, religion, ethnic identification, or residence. Larger external systems impinged on their lives only occasionally. The transformation and disruption of these local communities took much the same form as in other parts of the less developed world. Western medicine and hygiene eliminated plagues and the sharply reducing death rates that followed doubled and sometimes tripled the rate of population growth. Cities grew exponentially in size, creating a new urban milieu. Nation-states emerged through the visions of national leaders like Muhammad Ali in Egypt, Attaturk in Turkey, and Reza Shah in Iran. Primate cities in these new nation states intruded into the villages of the countryside as programs of health, education, and land reform were carried to even the most remote regions of the new road systems of the Middle East.[10] For the tribes, sedentarization (both enforced and voluntary) attended the submergence of the land of insolence beneath the technological superiority of urban governments. The passing of the <u>ecological trilogy</u> marks the emergence of a new human geography in the Middle East, one that will require new frameworks and new methodologies.

It is to this corpus of interdisciplinary inquiry into the nature of Middle Eastern society and economy that geographers have made their contributions. The broader formulations concerning city, village, and tribe in this region have been framed by several generations of archaeologists, historians, Orientalists, and a growing number of social scientists. Integration has occurred through the questions raised rather than by the special skills of the interrogator; generalizations have emerged through interchange among scholars of varied hue working on diverse places and periods. The remainder of this paper, then, is devoted

---

[9] A major theme in Janet Abu-Lughod, "Problems and Policy Implications of Middle Eastern Urbanization," Unpublished Ms.

[10] Salah el-Shakhs has argued that the high degree of primacy found in the Middle East represents an evolutionary stage in economic development. See his "Development, Primacy and Systems of Cities," <u>Journal of Developing Areas</u>, Vol. 7 (1972).

to a tentative discussion of the passing of the ecological trilogy in the Middle East. The writings of geographers are not treated apart.

The Middle Eastern City

Muslim cities were the cornerstones of Islamic culture and civilization, though urbanites never formed a majority of the population in the Middle East. Like cities elsewhere, Middle Eastern cities were centers of local economy, but the preeminence of cities was especially strong in this region because Muslim political, religious, and economic leaders virtually always lived in them. As a result, the major works of Islamic literature, philosophy and science and the more mundane systems of social and economic organization which were created in the cities of the Middle East were only casually influenced by the numerous but inarticulate peasantry in the region. The early Arab geographers defined their cities in rather specific terms. To them, a city or madina was any settlement which contained a Friday-prayer mosque and a permanent market place or bazaar. The designation of these two institutions as critical components of a city suggests the commercial and cultural leadership that Middle Eastern cities exercised over the region after the Muslim conquest.

Under Islam, the cities of the Middle East gradually evolved a distinctive flavor and coloring which tended to obscure their diverse histories, functions, and physical settings. These cities came to share distinctive internal spatial patterns marked by the allocation of central space to specific institutions, the development of characteristic patterns of social and economic organization, and the creation of an urban milieu congenial to the practices of Islam. Virtually every urban observer attributes these similarities in life and landscape to the Islamic cultural tradition, explaining similarities among the cities of the region in terms of Islam.[11] On this basis, Muslim cities have been distinguished from the urban centers of medieval Europe, from the cosmological centers of early China, and modern Western metropoli.[12] While no single Middle Eastern city

---

[11] See footnote 4 above, particularly von Grunebaum (1961). A similar view is held by Norton Ginsburg, "Urban Geography and 'Non-Western' Areas," in Philip M. Hauser and Leo F. Schnore (eds.), The Study of Urbanization (New York: Wiley, 1965), pp. 311-312.

[12] For an alternate view, see Gideon Sjoberg, The Preindustrial City: Past and Present (Glencoe: Free Press, 1960); idem, "Cities in Developing and in Industrial Societies: A Cross-Cultural Analysis," in Hauser and Schnore (1965), pp. 213-263.

rigidly exhibits all of these characteristics, the majority of Middle Eastern cities display general similarities in the organization of space, a central mosque, and bazaar complex, and residential space divided into quarters.

Limits of Urban Growth

At the beginning of the nineteenth century, cities in the Middle East were small, although a surprisingly high percentage of the region's total population, perhaps as much as a quarter, were town dwellers. Cairo, Baghdad, and Istanbul were the only cities in the region with populations larger than a quarter of a million although one out of every ten Egyptians lived in towns larger than 10,000 and similar figures were even higher for Iraq and Turkey. Cities as famous as Aleppo and Damascus were slightly larger than 100,000; Jerusalem and the future capital of Iran, Teheran, had populations of 10,000, respectively. Medieval urbanization had progressed by a proliferation of small urban centers rather than by growth and expansion of a few large cities.[13]

Several natural and social factors related to the functional role of cities in traditional Islamic society limited the size of each individual city. Although cities like Istanbul, Damascus, Baghdad, and Isfahan are best known to Westerners as international trading centers dealing in caravans of silk and spices, in fact, the international trading function rarely employed more than 10 per cent of the residents of these cities. For the most part, city-dwellers were engaged in producing, collecting, and processing raw materials from the hinterland--wool for carpets and shawls, grain and vegetables to feed the urban population, and nuts, dried fruit, hides, and spices for export.

Limited regional transportation networks reduced the size of the effective hinterland from which cities could reliably draw an agricultural surplus and thereby limited the size of urban populations and the growth of each urban center. In part, this constraint was relieved by the substantial numbers of urban dwellers in traditional Middle Eastern cities who were agriculturalists. As late as the beginning of the nineteenth century, over half of Cairo's adult males were employed as farmers or laborers.[14] Open spaces in the city were given over to cultivation, obviously alleviating the necessity of extensive external patterns of organization. What this suggests, of course, is that substantial portions of the

---

[13] Charles Issawi, "Economic Change and Urbanization in the Middle East," in Lapidus (1969), pp. 102-121.

[14] Abu-Lughod (1971), p. 59.

urban population had as much (or more) in common with the countryside as with the elites of the urban center. An ebb and flow of population away from the city in times of political crisis, plague, or depression and towards the center in better days suggests an adaptive fluidity of interaction that explains the wildly fluctuating urban population estimates one finds in the literature. Guild associations, juridical arrangements, land tenure contracts, and even religious orders spanned both types of communities integrating city and village into a broader regional ecosystem.[15]

To support the population of a major urban center like Istanbul in the preindustrial period required an organized network of regional centers, market towns, villages, and nomadic groups.[16] During the sixteenth century, regional centers of 20-40, 000 were rather evenly distributed over the territory of the Ottoman Empire, with many strategically located at the intersections of important trade routes or break-in-bulk points. These regional centers specialized in one or more types of production such as carpets, footwear, or cloth and were usually surrounded by belts of villages averaging about 500 households. The regional center produced goods for consumption in these villages and acted as central places for their immediate regions. The production activities of the regional center strongly influenced economic activities in surrounding market towns and villages, and the group of settlements relating to each center acted as a regional production unit linked to other such settlement webs and ultimately to the capital city. While nomadic groups were less well integrated into this hierarchy than more rooted communities, these tribesmen were an important source of income for the Ottoman state and when necessary were resettled and in some cases forced to become sedentary.

The success of this organized settlement hierarchy clearly depended upon safe and regular communications; this was provided by village societies called _derbents_ who maintained and repaired the roads and protected wayfarers from

---

[15] A growing theme in the anthropology literature. See Robert A. Fernea, "Gaps in the Ethnographic Literature on the Middle Eastern Village: A Classificatory Exploration," in Antoun and Harik (1972), pp. 75-102; John Gulick, "Village and City: Cultural Continuities in Twentieth Century Middle Eastern Cultures," in Lapidus (1969), pp. 122-158; Laura Nader, "Communication between Village and City in the Middle East," _Human Organization_, Vol. 24 (1965), pp. 18-24; P. Suzuki, "Encounters with Istanbul: Urban Peasants and Rural Peasants," _International Journal of Comparative Sociology_, Vol. 5 (1964).

[16] Ilhan Tekeli, "The Evolution of Spatial Organization in the Ottoman Empire and the Essentials of the Regional Policy of the Turkish Republic," forthcoming in Brown (1973).

marauding tribesmen and highwaymen. Speed of communication was assured by horseback messengers. Even more critical to the success of the system, however, was the continuing penetration of the vast area of Anatolia by the central administrative structure in order to maintain the delicate balance of local areas and regional units and ensure their continuing support of Istanbul.

But even given effective organization of territory by a centralized state, certain natural factors inside preindustrial Muslim cities limited their capacity to expand. When urban populations grew usually by migration from the countryside, problems of water distribution, sewage disposal, and food supply produced breeding grounds for malaria and cholera plagues causing high urban death rates. Fires and earthquakes also took their toll. In Istanbul, for example, 160 fires were recorded in a period of seven years, and the last of these, the great fire of 1865 reduced four-fifths of the city to ruins.[17] These physical and cultural conditions inhibited the growth of individual urban centers and encouraged urban proliferation.

Internal Patterning

The spatial patterning of these Middle Eastern cities until recently differed markedly from that of the industrial centers of the modern world. Several characteristics which consistently reappear in these cities, from Morocco to Afghanistan, are: (1) an emphasis on enclosure and security both in urban plans and domestic architecture, (2) the allocation of central space to a mosque-bazaar complex, and (3) organization of residential space into semi-autonomous quarters (ḥarāh) and streets (darb) usually with some degree of ethnic, occupational, or religious unity.

Every important Middle Eastern city (except Mecca, which is ringed by mountains), was surrounded by high walls and fortifications. Given the stormy history of the region, the expenditure of time and energy required to build and maintain these walls was a wise investment. The most common pattern was a circular or rectangular wall penetrated by four or five gates through which all contacts with the outer world filtered. The gates were usually the endpoints of major avenues which ran from a central mosque-bazaar complex to the walls, and were the only articulated means of circulation in the city. In Iranian and Afghan cities, still higher fortifications were built around a citadel inside the city walls. Often located on one side of the city, these higher walls protected

---

[17] Planhol (1959), p. 36.

the seat of local government both from internal uprisings and external attack.

This theme of enclosure and security was also evident in the organization of the residential quarters and individual houses in traditional Middle Eastern cities. In contrast with the general geometry of the urban plan, the organic, irregular pattern of these quarters is universal in the cities of the region.[18] The streets and lanes here form a maze of dark twisting passageways, alleys, and culs-de-sac. Most are less than twelve feet wide and are enclosed by the high outer walls of household compounds. Openings in the walls are heavily barred, their tops often studded with spikes or fragments of glass. There are no street signs and no numbers; there are no traffic jams because there is no traffic. Privacy, protection, and security were paramount in their construction.

Central space in these walled cities was allocated to two fundamentally Islamic institutions, the Friday-prayer mosque and the bazaar. The mosque was often the most beautiful and impressive building in the city and the pivot of social and religious life. Some like the al-Azhar Mosque in Cairo, Hagia Sophia in Istanbul, and the Masjidi Shah in Isfahan were also centers of major educational and administrative complexes. As cities evolved through time, new mosques became the focal points of new bazaars and the larger cities became polynucleated. Each mosque was a hub of activity for the religious elite, the 'ulamā', who applied Islamic law to family life, regulated markets and even water supply, and managed the city's religious, educational, and legal institutions.[19] The pervasive influence of Islamic values on the structure of Middle

---

[18]The maze-like character of residential quarters in Middle Eastern cities is one of the most striking visual similarities among them. This is usually attributed to factors in Islamic Law which mitigated penalties for encroachment on the public way. Only residents of a street were responsible for maintaining access, and unless the interference was sufficient to stimulate complaint, no remedial action was taken. Once standing, the interfering structure usually remained. See Robert Brunschvig, "Urbanisme médiéval et droit Musulman," Revue des études islamiques, Vol. 15 (1947), pp. 127-155. Also Abu-Lughod (1971), pp. 65-69; Planhol (1959), pp. 14-22.

[19]The administration of Middle Eastern cities is a complex and poorly understood subject. Two figures, the qāḍi and the muḥtasib, both religious appointees, held varying powers of urban administration at different periods in different cities. The qāḍi apparently primarily functioned to maintain order in the city; the muḥtasib in some cities regulated the water supply, maintained the streets, and regulated the bazaar, in others he functioned only as the administrator of the bazaar. More generally, the 'ulamā was quite influential in city affairs, and through the Islamic system of endowments (waqf) directly influenced urban land use in addition to their educational and legal authority.

Eastern cities probably derives from the paramount importance of these men in urban administration.

The second central institution in Middle Eastern cities, the bazaar, was a welter of crowded lanes, covered stalls, and associated caravanseries far noisier and more aromatic than Western financial districts. The bazaar was the locus of virtually all wholesale and retail trade within the city. In the bazaar, central location was highly valued. Thus, a concentric hierarchy of trades and crafts often existed with the shops of prestigious craftsmen such as booksellers, carpet merchants, jewelers, and silversmiths located at the center of the bazaar, the lesser crafts progressively nearer the margins. Even in Herat, which had a star-shaped central business district derived from the Hellenistic origin of the city, high-value shops were located at the center of the bazaar although distributors of frequently purchased, low value materials such as bread, candy, and spices were dispersed throughout the bazaar.[20]

The remaining space in traditional Middle Eastern cities was subdivided into residential quarters (ḥārah) which tended to be social as well as geographic entities whose people were bound together by ties of religion, blood, occupation, or common origin. During the Ottoman period, as noted earlier, this arrangement was formalized into the millet system which recognized the existence of a large number of separate and distinct social communities.[21] In Antioch, for example, forty-five such identifiable quarters existed with Jews, Armenians, Greeks, Europeans, and Muslim sects identified with specific sectors of the city. Each quarter formed a nearly independent community walled off from the others, a city within a city, with its own mosque or church, its own customs, law, and milieu.[22] At the turn of the nineteenth century, Cairo had a similar cellular residential pattern, although not demonstrating the same rigidity of segregation. The city was divided into fifty-three quarters, each of which consisted of several streets (darb). It was the darb, a unit of thirty or more house-

---

[20]English, forthcoming in Brown (1973).

[21]The millet system is discussed in H. A. R. Gibb and Harold Bowen, Islamic Society and the West (London: Oxford University Press, 1950), Vol. 2, pp. 207-261. See also Albert H. Hourani, Minorities in the Arab World (London: Oxford University Press, 1947).

[22]Jacques Weulersse, "Antioche, essai de géographie urbaine," Bulletin d'études orientales, Vol. 4 (1934), pp. 27-79; idem, "Antioche, un type de cité d'Islam," Congrès international de Géographie (Warsaw), 1937, Vol. 3, pp. 255-262.

holds, which was barricaded at night, which formed the learning environment of the child, and was the basic unit of urban social life.[23] In smaller cities like Kirman (except for the Zoroastrian Quarter) and Herat, rigidly walled quarters were absent.[24] Nonetheless, the same pattern of differentiation on the bases of race, religion, and common origin were primary axes of social organization, and small cohesive urban groupings were preeminent features of the social ecology of these traditional Middle Eastern cities.

Modern Forces of Change

This traditional Middle Eastern urban setting has been reshaped by the introduction of Western transportation and communication systems, by the development of new attitudes, values, and ideologies, and by the spreading availability of Western technology. Wide avenues suitable for motor transport have been built in most cities, and commercial activity has been liberated from the bazaar. Modern buildings have been constructed on these avenues--drugstores, gasoline stations, photography shops, hotels, and even department stores. Newly-built suburbs have spread beyond the confines of ancient city walls; the cemeteries and shrines which previously had acted as a collar on urban expansion have been levelled, providing striking evidence of the rising importance of secular considerations in a traditionally religious society. The urban elite are moving to these suburbs, where houses are less crowded and street patterns less tortuous. Another modern institution, the shanty-town or bidonville, has also emerged on the outskirts of large cities, pointing up the major growth of urban centers in modern times.

In the last century, the urban area of the city of Cairo grew from five square miles to seventy-five, making the old city a small district in the new metropolis. Tehran, the capital of Iran, grew in population from 100, 000 in 1940 to more than three million. Other cities like Beirut, Baghdad, Casablanca, and Istanbul experienced explosive urban growth. Although growth rates varied in tempo and magnitude from city to city, the factors underlying this urban implosion were two: first, the doubling of population region-wide caused primarily by declining death rates, and second, the immigration of peasants and tribesmen

---

[23] Abu-Lughod (1971), pp. 64-65.

[24] English, "Culture Change . . . ," (1966), pp. 34-35; <u>idem</u>, forthcoming in Brown (1973).

from the countryside to the expanding cities.[25] Roughly two-thirds of the current growth in these cities is occurring as a result of natural increase among urban residents; the drain of men from the countryside accounts for the remaining third.[26] Specific cities grew by dint of a combination of local factors--Amman, as a direct result of the Arab-Israeli war of 1948, Kuwait as a result of oil-derived industrialization. But region wide, the processes of modernization and urbanization reflected the integration of the region into larger networks of social and economic organization, the emergence of nation states as the most meaningful units of territorial organization, and a growing gap between the urban centers with access to the goods and services of Western technology and the village and nomadic camps of another era.

This integration of the Middle East into large-scale modern commercial systems led to a redistribution of urban populations in the region. Historically, Middle Eastern cities were international trading centers, connecting points on the inter-urban trade routes which criss-crossed the Old World. Located at the intersection of three continents and bordered by seas and deserts, these cities were important break-in-bulk points and staging areas for international trade flowing between Europe and the Orient and to a lesser extent between sub-Saharan Africa and the coasts of the Mediterranean. In the nineteenth century, the largest cities in the Middle East were located in the interior of the region. This interior location reflected the introverted nature of the cultural and economic life of the Middle East at that time, largely the result of European control of the Mediterranean Sea during the later Middle Ages. Many of these cities were caravan centers on the desert margins, like Damascus, Aleppo, Yazd, and Kirman; others were political and administrative centers, such as Cairo and Baghdad. Smaller cities were strung like beads along the caravan routes between these major centers, and prospered or perished depending on prevailing conditions of security, politics, water supply, and terrain at any given time. But when Western merchants and capital penetrated the region and began to integrate it into a wider, Western-oriented commercial world at the end of the nine-

---

[25] On population growth, see M. A. El-Badry, "Trends in the Components of Population Growth in the Arab Countries of the Middle East: A Survey of Present Information," Demography, Vol. 2 (1965), pp. 140-186; J. I. Clarke and W. B. Fisher (eds.), Populations of the Middle East and North Africa (New York: Africana, 1972); Gary L. Fowler, "Population Growth Trends in the Middle East and North Africa," Unpublished Ms.

[26] Abu-Lughod, Ms.

teenth century, this pattern of urban centers changed. Heterogenetic seaports like Beirut, Tripoli, Alexandria, Tangier, and Casablanca swelled in population and commercial importance. Road and railroad networks were built to connect these growing coastal cities with the hinterlands of the interior. Many of the great cities of the interior began to lose population as the seaports grew at the expense of the desert ports.[27]

Internally, these new cities are being restructured on new premises, not simply physically but socially as well. The traditional overlapping of family ties, religious affiliation, occupation, and place of residence is weakening. The Islamic social system based on ethnic division of labor, the "mosaic" which Carleton Coon traces back to early times, has been altered by the flood of change impacting the region after World War II.[28] Perhaps the most striking evidence of the impact of social change on spatial structure has been the abandonment of religious segregation as a principle of town planning. In Kirman, the Zoroastrian quarter disappeared in the growing atmosphere of religious tolerance (or increased secularism) in less than two decades.[29] A similar restructuring of the relationships between Muslims and Jews apparently occurred in Morocco.[30] What will replace the binding ties of the past is not at all clear. A detailed social area analysis of Cairo identified thirteen "sub-cities" and three social types (rural, traditional urban, and modern urban) which are in varying degrees converging towards the general configuration of a modern, Western, metropolitan class society.[31] What is clear is that a new urban milieu is rapidly replacing the traditional Islamic city.

Middle Eastern Villages

The second component of the ecological trilogy, the village, is the foundation of Middle Eastern society because from 70-80 per cent of the people in this

---

[27]Issawi in Lapidus (1969), pp. 108-111.

[28]Coon (1958), pp. 3-9.

[29]English, "Culture Change . . . ," (1966), pp. 40-41.

[30]Lawrence Rosen, "Muslim-Jewish Relations in a Moroccan City," International Journal of Middle East Studies, Vol. 3 (1972), pp. 435-449.

[31]Abu-Lughod (1971), pp. 182-220. This is the only city in the Middle East with a detailed analysis of modern structure. Clearly, this area deserves more attention from urban geographers.

region live in rural agrarian communities.[32] Occupied with cultivating wheat, barley, the date palm and a host of lesser crops, villagers form a majority in every major Middle Eastern country except Saudi Arabia and Jordan (where nomads hold sway) and Kuwait (where city and nation are virtually synonymous), and Israel (a Western economy). In general, villages are located wherever a reliable supply of water for cultivation is available, or can be made available through wells, horizontal tunnels (qanāt), water wheels, or other lifting devices. Particularly dense rural populations are found on the coasts of the Mediterranean, Black, and Caspian Seas, in the uplands of the Fertile Crescent where winter rains provide water, and in the river valleys of the region where irrigation systems have been constructed. After millennia of human experimentation, however, only the highest mountain regions and the cores of the great deserts of the Middle East--the Egyptian Sahara, the "Empty Quarter" of Saudi Arabia, the Syria Desert, and the kavīr and lūṭ regions of the Iranian plateau--have no settled rural population. In short, villages are ubiquitous in this region; in Iran, for example, there are 55, 000.

The Village Landscape

Most Middle Eastern villages are small compact clusters of dwellings with few facilities other than a mosque, a shrine, a few shops, and possibly a bath. The houses are often dome-roofed cubes built of adobe, mud bricks, or temped earth, which except for nearby fields and trees would blend indistinguishably into the surrounding landscape. Each house is surrounded by walls and oriented inward to the courtyard, creating even in small villages a network of alleyways which casts an illusion of urbanity. This housing pattern diversifies in those parts of the Middle East where alternative building materials, environmental constraints, and architectural traditions have come into play. In the marshes of southern Iraq, for example, Arabs live in large elaborate reed houses which along the western coast of Saudi Arabia take a beehive form (also, strangely enough, found near Aleppo). In the Saharan oases and in southern Saudi Arabia, palm fronds are a principal building material. In Libya, sandstone is used as adobe is father to the east.[33] In the highlands of Anatolia and

---

[32]Regional variations in occupational structure for each Middle Eastern country are summarized in the excellent set of tables in Nieuwenhuijze (1971), pp. 89-155.

[33]Robert Holz, "Sandstone as a Building Material in Libya," The Pennsylvania Geographer, Vol. 8 (1970), pp. 1-10.

Iran, wood and stone villages are interspersed with troglodyte settlements where villagers have solved the problems of construction and heating by living underground. In most cases, however, settlements are agglomerated; dispersed settlement is virtually non-existent in the Middle East.

The specific sites of these clustered settlements are usually carefully adapted to local environmental realities and to the daily patterns of village life. In areas where water is plentiful and cultivation is continuous, the cluster of houses (usually separated into village factions), mosque and shops is frequently located at the center of the village fields. Olive and date trees and cash crops are cultivated in walled enclosures, to foil adventurous animals, and form an inner ring around these structures at the core of the village where soil fertility is highest and security is greatest.[34] Grain fields usually radiate outward from the village providing maximum access to the cultivated area and minimizing the amount of daily walking required of each peasant, a factor of importance because of extreme fragmentation of land holdings. In Egypt, for example, the majority of fellahin work less than an acre of land (72 per cent own or work less than one feddan or 1.038 acres). In the largest villages of 15,000, the total cultivated area is about 1,500 acres and a trip from the village to the boundaries of the fields is less than a mile.[35] Where these conditions obtain, in, for example, the Nile delta and certain parts of the Levant, the distribution of villages on the landscape nearly approximates the hexagonal pattern hypothesized in Christaller's central place theory where the distribution of population is principally influenced by the friction of distance.

In other parts of the Middle East where the source of water is a stream, a canal, an aqueduct, or a qanāt, village houses are sometimes strung out along the watercourse and the village takes on a linear or triangular shape with the houses and walled gardens at the top or apex of the triangle and the grain fields fanning out downslope.[36] In these villages, the location of each household com-

---

[34]A tentative effort to apply the von Thünen model to the internal structure of Turkish villages is: William A. Mitchell, "Turkish Villages in Interior Anatolia and von Thünen's 'Isolated State': A Comparative Analysis," Middle East Journal, Vol. 25 (1971), pp. 355-369.

[35]Figures from Patai (1967), p. 272.

[36]Specifically dealt with in English, City and Village . . . (1966), pp. 50-54. Perhaps no single technological complex in this region has fascinated geographers as much as the qanāt. See George B. Cressey, "Qanats, Karez and Foggaras," Geographical Review, Vol. 48 (1958), pp. 27-44; Paul Ward English,

pound determines the quality and even the quantity of its water supply. One finds, therefore, the homes of prosperous landlords, the village shaykh and other distinguished villagers in the upper sections where water is clean and plentiful; the poorer households of peasants, sharecroppers and day-laborers are located down stream where the water has been polluted by use and has diminished in volume by evaporation, seepage, or even theft. Here, the administration of water becomes a critical social task, conflict resolution a vital human theme.[37] Where these conditions obtain, the distribution of villages on the landscape takes on quite different patterns, in irrigated lowlands strewn like beads along the watercourses and in mountainous regions taking on the tree-shaped pattern of the tributary valleys.

As is clear from above, village life in the Middle East is intimately tied to local environmental conditions. The vast majority of villagers are agriculturalists whose land provides the staple grain of his diet, whose few sheep and goats simultaneously provide protein and wool products for clothing, whose entire life is regulated by the rhythm of the agricultural cycle. Plowing and planting of the winter grains takes place in the fall, the harvest in late spring or early summer. The winter is a period of minimum activity when the villager scours surrounding wastelands for wood, works as a day-laborer repairing a well, or builds a house to accumulate that small amount of cash from which amenities like a mirror, a tea set, or a mat must spring. Basically self-sufficient, the villager's most important dealings are in kind. The local barber, potter, and carpenter are paid in grain at harvest time with the amount largely depending on the success of the harvest. The villager's few cash ventures take place in the town or city, which he visits infrequently. But the conditions of daily village life are strongly influenced by contacts with the city, and urban-village relationships are often at the heart of village organization.

"The Origin and Spread of Qanats in the Old World," Proceedings of the American Philosophical Society, Vol. 112 (1968), pp. 170-181; Johannes Humlum, "Underjordiske, Vandigskanaler: Kareze, Qanat, Foggara," Kulturgeografi, Vol. 16 (1965), pp. 81-132.

[37]Patterns of social and economic organization related to water use have formed an important basis for the study of the village and the nature of Middle Eastern society. See Robert A. Fernea, "Conflict in Irrigation," Comparative Studies in Society and History, Vol. 6 (1963), pp. 76-83; idem, Fernea (1969), pp. 118-149; and more broadly, Karl A. Wittfogel, Oriental Despotism (New Haven: Yale University Press, 1957).

City and Village

The primary relationship between cities and villages in the Middle East is of course economic, since cities rely on peasants for their food supply and village farmers must market a portion of their crops. In Turkey, where three-quarters of the land is farmed by peasant owners, this exchange is accomplished through a stratified settlement hierarchy. Significant variations occur at each level of the hierarchy depending on location within the system, with some villages being "urban-directed," others "rural-directed."[38] In parts of North Africa, tribal farming groups have family components living in both cities and villages, an arrangement leading to direct exchange without intermediate marketing. Until very recently, however, the vast majority of cultivated land in the Middle East was owned by urban landlords, and land tenure agreements between these landlords and village sharecroppers formed a basic economic pattern. While the terms of these contracts vary, in theory, the harvest is divided into five portions (khamseh in Iran, khammes in North Africa). Each portion is allocated to the party which supplies one of the five elements of production--land, water, plow animals, seeds for planting, and labor.[39] The peasant always receives one-fifth of the harvest for his labor and more if he provides either animals or seeds. In a good year, the peasant is able to provide his family with sufficient food and a bit besides. But in lean years, which are frequent because the environmental risks of drought and earthquake and the social risks of disruption and epidemic are high, the peasant is instantly in debt, lacking sufficient economic resources to cope with even minor disturbances.

As a result, the landlord generally is expected to protect and guard the villager, insofar as possible, from the threats of government tax-collectors, marauders, and money-lenders. More subtle provisions of these land tenure contacts deal with the provision of fertilizer (if any), whether payments to the village mosque and carpenter should be taken out prior to the division of the

---

[38]The complete typology offered by Kolars includes seven types, four of which are "urban-directed" (shadow, annexed, satellite, and summer dormitory) and three are "rural-directed" (market-seeking, market-recognizing, and market-ignoring). Kolars in Shorter (1967), pp. 63-83.

[39]Ann K. S. Lambton, Landlord and Peasant in Persia (London: Oxford University Press, 1953), pp. 306-329. Hans Bobek uses the term "rent capitalism" to describe this system. See his "The Main Stages in Socioeconomic Evolution from a Geographic Point of View," in Philip L. Wagner and Marvin W. Mikesell (eds.), Readings in Cultural Geography (Chicago: University of Chicago Press, 1962), pp. 218-247.

crop or later, whether the stubble in the field can be grazed by the villager's animals, whether the straw belongs to landlord or peasant. On these seemingly trivial considerations, the health of the family and the success of an agricultural year often hinges.

Before recent land reform programs, absentee ownership of village land was a general pattern in the Middle East, but wide variation in city-village relationships was produced by the obvious factors of location, economics, and terrain, and by more subtle forces derived from the history of settlement, corporate as opposed to individual ownership of land, and the tribal or kinship basis of village society.[40] In Kirman, for example, the sharecropping system acts as a mechanism of urban dominance which annually drains economic surpluses from surrounding villages, leaving the mass of the rural population quite literally without resources. Similar contracts in herding and weaving reinforce urban control of urban resources, leaving village life shaped by social and cultural patterns emanating from the city as well as economic directives.[41] In the highland regions east and west of the Jordan and in other parts of the Levant are communal villages (musha'). Here, each male head of household owns a share of village lands and these parcels are redistributed periodically within the village. In southern Iraq, large tribal confederations with both nomadic and sedentary sections resisted urban domination for long periods of time depending on communally owned resources and cooperative maintenance of irrigation systems for their livelihood.[42] A more autonomous and individualistic pattern may apply in villages distant from major urban centers in areas where sufficient rainfall requires little or no cooperative development.

Forces of Change

In modern times, the bonds between city and village have intensified dramatically as political regimes in the new nation states of the Middle East have attempted to bring their countries into modern twentieth-century life. A revolution in transportation and communication systems is gradually carrying social

---

[40] An excellent typology of Middle Eastern village based on two criteria, tribal organization and degree of urban domination, is provided in R. A. Fernea, Ms.

[41] English, City and Village . . . (1966), passim.

[42] R. A. Fernea (1969), pp. 38-54.

and economic ferment from the government bureaus and bazaars of the city into the countryside.[43] With increased access, peasants are being integrated into larger national social and economic systems. The agents of change are several: the land reform agent who supervises the distribution of land to former sharecroppers, the government teacher convinced of the value of literacy and occasionally the equality of women, the doctor who teaches health and sanitation. The actual processes of diffusion have never been studied, but modern technology in the form of tractors, bicycles, radios, sewing machines, and hurricane lamps is already available in many areas.[44] Nevertheless the basic standard of living in the Middle Eastern village has not greatly changed; insecurity is still a chronic condition.

It is for this reason that when modern writers speak of social conditions in the Middle East--of poor health, of high infant mortality rates, of illiteracy, tenancy, and numbing poverty--they are primarily describing conditions in the Middle Eastern village. Diets are insufficient and heavily weighted towards carbohydrates, so that protein deficiencies and malnutrition are common in Egypt, the Fertile Crescent, and Iran.[45] Although malaria is largely under control, trachoma, dysentery, and bilharzia are still widespread. The cumulative effect of these conditions produces infant mortality rates in some parts of the Middle East such that one of every two children dies before reaching the age of one year.

But if modernization has not significantly raised standards of living, it has disrupted the traditional matrix of the Middle Eastern village. New crops, improved agricultural technology, and new techniques of cultivation are modifying the agricultural basis of village life. The traditional reverence for age and religion is decreasing as power flows into the hands of young, educated representatives of central governments, and as religious leaders are shorn of their control over communication, education, and law. Whereas in the past, the unrelenting economic insecurity of peasants in the Middle East found some amelioration in the comfort and satisfaction of life in a small, coherent community, these changes

---

[43] See, for example: John Kolars and Henry J. Malin, "Population and Accessibility: An Analysis of Turkish Railroad," Geographical Review, Vol. 60 (1970), pp. 229-246.

[44] One of the only attempts to study the diffusion of a trait as an index to culture change in the Middle East is Brian W. Beeley, "The Turkish Village Coffee House as a Social Institution," Geographical Review, Vol. 60 (1970), pp. 475-493.

[45] See tables in van Nieuwenhuijze (1971), pp. 104-105, 152-155.

are causing the young of the village to view their families, communities, religion, and nations in a new way. As this new vision prevails it will alter the social roots of the Middle Eastern village, which until recently has been the most stable, unchanging component of the Middle Eastern ecological trilogy.

Nomads of the Middle East

To use Carleton Coon's deft phrase, "nomads, like cowboys, are more glamorous than numerous"; they comprise less than 10 per cent of the Middle Eastern population, some fifteen million of an estimated two hundred million people. Only four major Middle Eastern nations, Turkey, Iran, Saudi Arabia, and the Sudan, have more than a million nomads; only in Saudi Arabia and possibly Jordan do they form a majority of the population. The domains of the nomads include the habitable portions of the deserts and steppes of the Sahara, the Arabian Peninsula, and the Levant as well as the mountain and plateau region of Turkey, Iran, and Afghanistan. The largest tribes in the Middle East are the Tuareg, Teda, and Chaamba of North Africa, the Bedouin of the Arabian, Syrian, and Egyptian Deserts, and the Kurds, Qashqa'i Bakhtiari, and Baluch of the mountains and plateaus of Anatolia and Iran.[46]

Nomadism

In general, three broad types of nomadism can be identified in the Middle East: (1) horizontal or "true nomadism" which involves the movement of herds of animals over large distances among the oases of the deserts, (2) vertical nomadism or trans-humance which describes the seasonal oscillation of herds of sheep and goats between lowland areas and mountain pastures, and (3) semi-nomadism on the steppe margins of settled agricultural communities usually involving at least some cultivation. All of these nomads live in impermanent dwellings such as felt yurts, camel-hair tents, or brush huts for at least part of the year. All follow regular patterns of movement adapted to the needs of the herds which form the core of their subsistence economy. Each specific type of nomadism is a highly rational adaption of human life to local resources which enables a substantial number of people and animals to subsist in severely restricted environments.[47]

---

[46] See footnote 7 above.

[47] A discussion of various classifications of nomadism in Asia and North Africa can be found in Johnson (1969), pp. 12-19.

Horizontal camel nomadism is practiced in the deserts and on the steppe margins of the Saharan Desert of North Africa and the twin desert cores of the Arabian Peninsular, the Rub' al Khali in the south and the Nafud in the north. These regions have extremely thin and scattered vegetation covers, partly for climatic reasons, partly through overgrazing. Fewer plant species exist here than in any region excepting the polar zone, and most plants are succulents or thorny bushes, low in nutritive value. Rainfall is so infrequent and irregular that houses in some desert oases are built of salt (without fear of melting) and some meteorological stations have never recorded any precipitation. A rainy year usually amounts to two or three showers. Given these conditions, animals kept in one place would soon eat themselves out of pasture, so that pastoral migrations take the form of movements in search of pasturage for the herds of camels.

Quite sensibly, desert nomads stay out of the desert as much as possible, setting up permanent camps on the steppe margins or in oases during the hottest and dryest season. As a result, camel nomads in the Sahara spend most of the year on the desert fringes; in the north on the steppes of Cyrenaica, Southern Algeria, southern Morocco, and Mauritania and in the southern Sahara on the edge of the Sahelian zone.[48] When rainfall comes, the nomads move out from these bases into the desert in search of pasture, and the long distance seasonal migration which may cover as much as a thousand miles and last as long as nine months begins. On the northern margins of the Sahara, the migration takes advantage of the winter maximum of such precipitation as exists, in the southern margins on the summer maximum. In Arabia, the great circuit of annual grazing begins in autumn and lasts through early spring.

Despite the camel nomad's heavy dependence on his herds, every tribe has access to the products of cultivation. Rarely tilling the land themselves, nomadic tribes control at least part of harvest in desert oases and steppe settlements such that in the Sahara; for example, only the peasants of the Souf and Mzab oases have been able to free themselves completely from all obligations to the camel-herding tribes.[49] Commonly, the nomads directly own from a quarter to a third of the date groves in desert oases and dates form a critical element in their diet. In the southern Sahara, the palm trees are tended by a slave caste of black farmers, the Haratin, who for all practical purposes are owned by the Tuareg

---

[48]Capot-Rey in UNESCO (1962), pp. 301-310.

[49]Sweet (1965), pp. 1133-1139.

and the Teda. The division of harvest is hardly equitable, for the tribes receive anywhere from one-half to four-fifths of the harvest, as among the Ahaggar Tuareg. In areas where the tribes do not directly own cultivated land, they are paid tribute, the "brotherhood tax," by the sedentary farmers for protection of the oases based on written agreements, some of which are centuries old. In both cases, the camel nomads return to the oases during the date harvest to collect their share, a major event which breaks the continuity of pastoral migrations in the Sahara. Similarly, in Arabia where the Bedouin control or own desert pasturages and wells, the smaller oases and even some large ones, oasis cultivators are slaves, clients, or tenants of the camel nomads. Subsidiary or vassal tribes of sheep and goat nomads on the fringes of the desert are also influenced by the Bedouin. Until the introduction of Western technology, in the Middle East, the desert ecosystem was largely controlled by camel nomads.

The ecological basis of <u>vertical nomadism or transhumance</u>, in the mountain and plateau areas of Turkey, Iran, and Afghanistan, rests on the existence of two kinds of seasonal pastures not suitable for year-round grazing or permanent cultivation--summer pastures in the higher hill and mountain ranges which are covered by snow in winter and winter pastures in the lowlands which are burnt off by the searing summer heat.[50] The utilization of these pastures through transhumance enables people and herds living in one ecological zone to utilize nearby zones during their period of a maximum productivity. If the herds were kept in a single zone year-round, approximately 70-80 per cent of the animals would die.[51] Variations in this type of nomadic existence range from transhumant movements of less than ten miles with residence in tents for less than three months to long distance migrations from one zone to the other covering 100 to 1,000 miles, implying almost continuous migration.

Frequently a succession of occupants with different flocks utilize the same pastures at different seasons. In the Zagros foothills of Kurdistan, for example, Kurdish sheep herders graze their flocks in the lowlands during the winter. As they migrate to higher elevations in the summer, Arab camel herders move in from the desert onto the vacated pastures, too parched for the sheep but adequate for camels. In the mountains of southwestern Iran, certain strains of sheep have greater resistance than others to changes in temperature or parched food,

---

[50] Johnson differentiates between "constricted oscillatory vertical nomadism" and "limited amplitude vertical nomadism." Johnson (1969), pp. 170-175.

[51] Estimate by Barth in UNESCO (1962), p. 373.

leading to similar patterns of sequent utilization. These differences among animals and even within a single species lead to specialization and to temporal successions which facilitate full utilization of the pasture resources of the region.

In Iran where transhumant tribes living in the Zagros and Elburz Mountains on the perimeter of the Central Plateau own a large proportion of the country's livestock, full utilization of pasture resources has led to overgrazing and destruction of plant resources.[52] The minimum size flock which permits survival for a transhumant family is approximately forty sheep or goats.[53] The principal hedge against disaster is to maintain as many animals as possible, leading to large scale overgrazing of forage resources. The primary threat to the nomad is not drought but rather cold spells during lambing time or at a crucial point in the migration which can destroy as much as 50 per cent of a herd overnight. Maintenance of large herds is a mechanism tribesmen use to avoid involuntary sedentarization in the face of these climatically induced cyclical alternations of prosperity and disaster. In spite of these hazards, most nomads perceive village life as a poorer, less satisfactory way of life.

The third type of nomadism found in the Middle East, semi-nomadism, encompasses a wide variety of agricultural and herding combinations which represent efforts to find the optimum economic fit to local environmental conditions. Solutions vary from year to year in response to rainfall regimes, social and political conditions, and in some cases market factors. Along the Euphrates, short-distance sheep and goat nomadism is practiced to take advantage of pastures in surrounding uplands. Occasionally a tribe will tend herds belonging to a complex of villages, but this system is unusual because for the most part the milk produced by village animals must be consumed immediately, meaning that

---

[52]Human impact on the Middle Eastern environment through charcoal burning and overgrazing has been a theme of constant interest to geographers. See, for example, John F. Kolars, "Locational Aspects of Cultural Ecology: The Case of the Goat in Non-Western Agriculture," Geographical Review, Vol. 56 (1966), pp. 577-584; Marvin W. Mikesell, "Deforestation in Northern Morocco," Science, Vol. 132 (1960), pp. 441-448; idem, "The Deforestation of Mount Lebanon," Geographical Review, Vol. 59 (1969), pp. 1-28; Hermann von Wissmann et al., "On the Role of Nature and Man in Changing the Face of the Dry Belt of Asia," in William L. Thomas (ed.), Man's Role in Changing the Face of the Earth (Chicago: University of Chicago Press, 1956), pp. 278-303; Gilbert F. White (ed.), The Future of Arid Lands (Washington, D.C.: American Association for the Advancement of Science, 1956).

[53]Stauffer (1965), p. 291. Barth (1961), estimates a figure of sixty.

their flocks are rarely more than one-half day's distance away. In North Africa, cultivation of the steppe margins by sections of nomadic tribes is common. Even among the mountain nomads of Iran and Afghanistan, villages are peopled by sedentarized nomads who will, on occasion, revert to nomadism should conditions allow. The flexibility of this system points up the keen environmental sensitivity of the peoples who occupy the frontier regions between the desert and the sown.

The Desert and the Sown

Historically, the frontier between the desert and the sown has been the locus of changing social forces and institutions in the Middle East, a boundary of sufficient flexibility and instability as to deeply influence the nature of Middle Eastern society.[54] Conflict between nomadic and sedentary peoples is stamped on the *Avesta* of the Zoroastrians, the early literature of Mesopotamia, the writings of Ibn Khaldun, and the culture and consciousness of Middle Eastern people. The permeability of the eastern and southern flanks of the region to infiltration and invasion by nomadic peoples frequently led to rearrangements of the ecological basis of livelihood. In Iran, for example, the current Pahlavi dynasty is the first in one thousand years not derived from a tribal incursion. Movements by Indo-Europeans, Arabs, Mongols, and Turks delineate the cultural subregions of the Middle East and reflect the outcomes of contests between tribal and urban based powers. The interfaces between settled peasantries and nomadic pastoralists form frontiers of culture contact critical to comprehending the nature of traditional and modern society in the Middle East.[55]

The complexity of relationships between agriculturalists and nomads seems to suggest that these groupings form interdependent and transferable forms of response to prevailing environmental uncertainty. In Mesopotamia, for example, the unpredictable nature of water supplies for irrigation combined with a hot, dry climate has led to a willingness on the part of cultivators to move along the continuum to pastoralism (and vice versa) in certain circumstances. Technological considerations in Mesopotamia restrict the zone of cultivation to riverine areas, leaving a vast expanse of pasture lands available for complementary and non-

---

[54] A central thesis in Robert McC. Adams, "The Mesopotamian Social Landscape: A View from the Frontier," unpublished *Ms.*

[55] Ian R. Manners, "The Desert and the Sown: An Ecological Appraisal," in Alice Taylor (ed.), *Focus on the Middle East* (New York: Praeger, 1971), pp. 27-38.

competitive use by pastoralists. Robert McC. Adams has argued that maintenance of a minimum standard of subsistence in the face of uncertain harvests and undependable water supplies has placed a premium on social institutions that favor a pooling of resources and the sharing of risks.[56] From this perspective, large tribal and clan groupings with components periodically shifting from agriculture to herding and therefore covering a range of environmental niches represent a pattern of social adaptation to environmental risk, which in turn explains the persistence and vitality of tribal organization in this region.

Similar patterns of adaptation explain the history of settlement in the Jabal al-Akhḍar region of Cyrenaica and in parts of the Middle Atlas region farther to the west.[57] Occupied by sedentary, subsistence-oriented agriculturalists during the Greek, Roman, and Byzantine periods, the region was occupied by Bedouin pastoralists after the Hilali invasions of the twelfth century. The rural population gradually absorbed a Bedouin system of social and economic organization and urban centers virtually disappeared. A pattern of ecological adaptation replaced the urban network which enabled these pastoralists cum agriculturalists to be virtually self-sufficient. The Jabal al-Akhḍar was divided into north-south strips controlled by separate lineage segments running from the Mediterranean, through the dissected plateau, downslope to the desert interior. This arrangement provided full access to all ecological variations and resources existing in the region, while allowing the large tribes to control key points along the ecological gradient. Cattle were herded along the coast, sheep and goats in the uplands, and camels in the interior. Barley and wheat especially were grown on the higher tablelands but also wherever conditions allowed along the coasts, in the beds of wadis, and on the fringes of the steppes; date palms were cultivated in the interior oases. Failure in one part of the system was compensated by a surplus elsewhere. Environment risk was dispersed, because mutual cooperation and reciprocal assistance within the tribe was expected. Thus trade amounted to internal family exchanges, periodic markets replaced towns, and critical functions were performed through patterns of tribal

---

[56] Adams, _Ms_.

[57] See Douglas L. Johnson, _Jabal al-Akhḍar, Cyrenaica: An Historical Geography of Settlement and Livelihood_, University of Chicago, Department of Geography Research Paper No. 148 (1973). See also Emrys L. Peters, "The Proliferation of Segments in the Lineage of the Bedouin of Cyrenaica," _Journal of the Royal Anthropological Institute_, Vol. 90 (1960), pp. 29-53 and "Some Structural Aspects of the Feud among Camel-Herding Bedouin of Cyrenaica," _Africa_, Vol. 37 (1967), pp. 261-282.

cooperation adapted to environmental conditions rather than by an organized settlement hierarchy.

A final example of the complexity and fluidity of interactions between sedentary and nomadic groups in the Middle East lies in the mountain and plateau regions of Anatolia, Iran, and Afghanistan. The transhumant nomadism practiced by tribes like the Qashqa'i, Bakhtiari, and Baluch presupposes the existence of permanent agricultural settlements and reciprocal exchange of products in a larger regional economy.[58] In some areas, tribesmen trade with larger markets through "village friends," agents who negotiate exchanges of pastoral products and equivalent village and urban products. Where nomadic tribes dominate, this exchange is effected through tribute or rent; when the power roles are reversed, however, nomads temporarily become agricultural tenants of sedentary landowners. As noted above, the level of risk in animal herding in these mountainous regions is so high that although successful nomads may purchase land and settle out as landowners, the more common route to sedentary life is through disaster, the more common role that of agricultural laborer. These on-going processes blur the distinction between herder and cultivator, between tribesmen and villagers, to such an extent that the categories themselves may tell us little about the nature of rural life in the Middle East. But it is obvious that nomads everywhere are part of a larger economic system, and that current changes in their numbers and adaptive patterns have direct consequences on sedentary components in large national or international systems.

Forces of Changes

In spite of the tribesmen's direct contribution to the economy and society of the Middle East, modernization has disrupted their culture, destroyed their independence, and in some cases impressed them into lives of enforced sedentarization. As long as both the governments of the Middle East and tribal confederations had the same technology--that is, horses, camels, and rifles--equilibrium was maintained. But recently the balance shifted so drastically as to threaten the place of tribal nomads in the new nation states of the Middle East. Airplanes and trucks conquered the deserts and mountains of the regions; nomadic raids provoke massive response by government troops armed with machine guns, tanks, and airplanes.[59] National boundaries truncate traditional

---

[58] Barth in UNESCO (1962), pp. 344-347.

[59] In Coon's view, "anti-nomad" technology. See his comments in S. N. Fisher (1955), *passim*.

migration paths. Sedentary groups pay "brotherhood taxes" to central governments, not to the tribes. In some countries at least, urban hostility to tribal groups influences the new leaders of Middle Eastern societies to accept major losses in livestock and pastoral products in return for the assertion of central dominance over the peripheral regions held by nomads.

In the Sahara, the ecological balance between nomadic peoples and the desert environment was upset by political and economic changes associated with European penetration of the region. Tribal pastures were brought under government control by the construction of army posts throughout the region under a program of "pacification." Bitterly resented, powerful tribes led by Islamic dervishes called marabouts, fought back, lost, and chronicled warfare in the Sahara ended. The new security imposed on the desert by Western forces encouraged cultivators to extend their fields into pastures formerly grazed by nomadic herds. On the steppe margins of the desert in Algeria, Tunisia, and Morocco, the Haratin "slave" caste was freed from tribal control and nomad and peasant alike paid taxes to the state. Simultaneously, the nomads suffered a disastrous loss in income as it became clear that caravan transport could not compete with trucks. Constricted in space, reduced in prestige, lowered in income, and led now by government appointed chieftains, Teda, Tuareg, and Berber tribes began to settle out in villages on the steppes, at wells in the desert, and in the emerging cities along the Mediterranean coast. Similar processes occurred in the mountain and plateau regions of Anatolia and Iran, where transhumant tribes have been disarmed and economic and political security has been established over their domains. Throughout the Middle East, then, tribal sedentarization is both general and irreversible. There appears to be no political place for tribal nomads in the new nation states of the Middle East.

Conclusion

The _ecological trilogy_ of the Middle East calls attention to sets of questions of deep concern to geographers, both by training and temperament. The interacting nature of city, village, and tribe in systems of spatial and social organization, the ecological basis of life forms in the region, the role of environmental uncertainty in social adaptation, the impact of new transportation and communication systems in delineating emerging systems of settlement, and on-going processes of landscape change are fundamental to comprehending current patterns of land and life in the Middle East. Certain clues suggest that an entirely new cultural and social geography is emerging in this region--the quantitative and qualitative growth in urban life and urban power, the secularization of critical social classes, the destruction of pastoral tribes, and the integration of village communities into national societies. How should geographers respond to these processes?

It is probably arrogant for any one discipline to suggest that its role is critical to understanding and/or solving the problems of a vast and diverse region, poorly known and partly studied. But risking that, several suggestions can be made. The varieties of new models and methods in the discipline, mainly quantitative but also non-positivistic, have never been applied in this region. As data bases emerge for cities in most countries, and hinterlands in a few, the possibilities for rigorous, formal analysis expand. To the best of my knowledge, only one city has undergone social area analysis, only one central place study is near completion, and no formal diffusion paper has been attempted. Studies of perception, decision environments, and non-Western environmental sets are hindered by the inability or unwillingness of most geographers to control language, as well as the common problems of research funding and field time. Second, virtually no studies exist which integrate social and economic components on a regional basis. Most research either is too general (like this paper) or site-specific like the anthropological ethnographies. It would seem that the opportunity exists for geographers to seriously impact conceptualizations of both the historical Middle East and the modern region. It would however be improper to truly believe that these new methods will expose the skeleton of human or spatial organization in the Middle East in any final way. This region by its diversity, complexity, and scale has eluded the generalizations of scholars for some time. It quite likely will continue to do so, which in turn will attract geographers to this study.

CHAPTER 8

AFRICAN GEOGRAPHICAL STUDIES AND COMPARATIVE REGIONAL DEVELOPMENT

Edward W. Soja
University of California, Los Angeles

Anyone charged with the task of assessing the contribution of geography to area studies must immediately contend with some vexing questions. First of all, why is the subject being raised? Surely not simply because no such assessment yet exists or because a concern for areas and cultures--must we define it as foreign areas and foreign cultures?--is built into geography. Furthermore, how can one evaluate a whole discipline's contribution to an understanding of a particular "cultural realm," especially one so wideranging as geography?

Obviously, if the task is to be attempted at all, one must be selective. But even after narrowing the focus to a set of themes and approaches which comfortably parallel one's own interests, where do you begin? As a geographer whose primary concern is with urban and regional development in Africa, do I expound on the perceptive tracings of Ibn Khaldun? Do I include a section on "Everything You Wanted to Know about Nairobi but Were Afraid to Ask"? Is Walter Christaller a major contributor to African geographical studies because of the growing use of central place theory in the interpretation of African spatial organization and in regional planning?

Extending this contemplative exercise still further, one can perhaps perceive a hidden agenda, for it is difficult to accept the need for still another compendium _cum_ bibliography of what geographers have written about Africa or Latin America, even if it should include some incisive commentary and criticism. But here again the possibilities are almost unlimited. In my most positive moments, I can recognize an opportunity for a group of experienced geographers to explore and evaluate cross-cultural and comparative research with an eye towards enriching the conceptual development of geography as a whole and enhancing the efforts to understand and solve the problems which exist in a given corner of the world. At my most cynical, I can also imagine a backslapping session bent on revalidating the credentials of traditional regional geography, punctured with satisfied sniggers at the absence of land use zonation around Kuala Lumpur or hexagonal trade areas around even the most isotropic surfaces of the Argentine Pampa.

It seems clear, therefore, that before beginning to examine the particular contributions of geographers to African area studies and the impact of African geographical research on geography as a whole, some preliminary statement of purpose, focus, and inclination must be made. What follows then is a brief outline of the assumptions, biasses, and personal preconceptions which underlie the approach taken in this paper.

Geography and Area Studies: A Subjective Viewpoint

Implicit in the conjunction of area studies and geography is the need to reexamine the nature and focus of regional geography. This reexamination, however, need not and indeed should not be bound to the older controversies which have characterized the role of regional geography in the past. However one might react to the changes which have taken place within geography over the past twenty years, there is no doubt that the discipline has been significantly if not completely transformed. And it is in this new context that the role of regional geography must be evaluated.

Regional geography can be defined as the comprehensive and integrative study of human spatial organization and behavior within a specified area of the world. As such, it is not only a necessary form of geographical specialization but a particularly exciting frontier for contemporary research. This excitement derives largely from an extension and synthesis within a more grounded, empirical context of the new ideas, models, methodology, and approaches which have emerged within modern geographical analysis--a process which has only just begun.

Certain more specific research objectives seem particularly noteworthy. These include (a) the testing and, if necessary, reevaluation of existing geographical theory in non-Western contexts; (b) the development of an explicitly formulated <u>comparative</u> framework for cross-regional analysis which would encourage greater interaction between different groups of area specialists; and (c) a concentrated attempt to focus research on those problems identified by foreign governments or specialists as affecting the welfare of their populations. These objectives can be summarized in the need to develop, as central to regional geography as a whole, a more theoretically rooted and practically focused form of comparative regional analysis.

This goal is not likely to be more easily attainable by working within the framework of area studies as they have conventionally been defined. Looking back over the growth of formal area studies in the United States, it is not difficult to conclude that, whatever its specific and occasionally interdisciplinary accomplishments, area studies specialization created another form of rigid compartmentalization within the social sciences. Area specialists often found themselves, in many cases by choice, isolated from other members of their own disciplines, including other area specialists, and united not by any intellectual focus or design but by a common emotional attachment and shared desire to live and work in a particular foreign area. The result was a peculiar kind of parochialism which inhibited the infusion of new ideas from their associated disciplinary origins and hindered effective cross-fertilization and comparative analyses among groups of area specialists.

Several factors reinforced this parochial isolation. The ethnographic model, for example, reflecting the important role of anthropology in the early growth of area studies specializations, led to a widespread acceptance of the need to spend many intensive years "in the field" before any valid interpretation could be made of foreign cultures. Great emphasis was placed on language learning and field techniques. Given the investment in time and effort necessary for this kind of training, it is no wonder that observations made by non-regional specialists or those without the prerequisite credentials were often automatically looked upon as naive, premature, and essentially irrelevant. This posture--often leading to what might be called the "what can he possibly know about the *real* Gabon" syndrome--frequently worked to discourage comparative research or at least to postpone it until all the world's people and regions had been accurately and fully described.

Perhaps even more influential was the ease with which area specialization could act as a refuge for mediocrity. African geographical studies, for example, could often be done without critical evaluation by other non-Africanist geographers. And if necessary, one could always shift audiences whenever convenient. Other disciplinary specialists in the same region could be viewed as not having the necessary background and skills of the geographer, while other geographers could be dismissed as not knowing enough about the "real Gabon." Such academic broken field running exists within disciplines as well, but seems to reach a higher level in area studies.

It is equally unlikely that traditional regional geography or regional science as it is currently practiced will act as progenitors of an acceptable conceptual framework for comparative regional analysis. The former is too incohesive and lacking in conceptual focus. Oddly enough, it also has not fostered a strong comparative perspective. True, there are the odd jewels of regional description we all proffer when our attachments to some form of regional geography are overzealously challenged. But what these usually inspire is duplication rather than comparison, and we all cannot be Vidal de la Blache or Henry James. Regional science, on the other hand, appears pointed in the right direction, but, when viewed with respect to comparative regional analysis, it seems sidetracked in an overemphasis on deductivist methodology, equilibrium states which never seem to exist in reality, economic as opposed to alternative variables, and spatial models which too often neglect space.

What seems most essential to the development of comparative regional analysis is perhaps most obvious: the need to achieve a balance between the particular and the general, between the empirical and the abstract, between a sensitivity to cultural differences and the ability to recognize uniformities in human spatial organization and behavior. Healthy skepticism or at least thoughtful hesitation should be maintained with regard to the applicability of abstract geographical models and theory, for the essence of a cross-cultural perspective is a deep respect for cultural relativism and the degree to which ethnocentrism can flavor even the most "objective" generalizations. At the same time, however, one cannot automatically dismiss existing theory when it does not initially appear to be appropriate in a particular area. More of lasting value is likely to emerge from even a slightly "forced" attempt to assess the potential applicability of general theory to a particular area than from the amassment of specific detail just to illustrate its irrelevance.[1]

Where then is the conceptual basis for comparative regional analysis likely to be found? Assuming a personal bias toward spatial organization as opposed to man-environment relations as the central focus of geographical study,

---

[1]The work of G. W. Skinner on rural marketing in China is a good example of what is meant here. Although both the central place theorist and the traditional China expert might recoil at Skinner's use of central place theory, his work represents one of the most insightful empirical applications of central place concepts in the literature. It is enriching both in an understanding of rural China and in its broader, conceptual implications. See G. William Skinner, "Marketing and Social Structure in Rural China," Journal of Asian Studies, Vol. 24 (1964-1965), pp. 3-43, 195-228, 363-399.

the most fertile source would appear to be the conceptual and methodological developments which have already taken place within what has been called the "locational analysis" school of geography. Of particular importance are the relatively few successful attempts at synthesis, such as those made by Berry and Haggett. Indeed, Haggett's "Locational Analysis in Human Geography" provides an excellent structural outline for comparative regional analysis and I have used it with interesting results to teach a course on Africa.

A conceptual framework resting solely on the foundation of general locational analysis, however, suffers from an overemphasis on form and static organizational structures. To it must be added a more dynamic approach to regional analysis and a more direct orientation to practical problems. These essential ingredients appear most clearly and cogently in the growing literature on regional development, ranging from the perceptive syntheses of John Friedmann, Tormod Hermansen, and E. A. J. Johnson to the more specific analytical studies of Brian Berry and Paul Ove Pedersen.[2] These works highlight the general processes affecting the organization of space, particularly the intertwined processes of urban systems development and spatial innovation diffusion, and provide both a dynamic theoretical focus for comparative regional analysis and the foundation for improved urban and regional planning.

In the remainder of this paper, an attempt is made to assess recent geographical research on Africa with specific regard to the development of a more theoretically based practical and comparative approach to regional analysis. There will be little attention given to the uniqueness of Africa, to the growing storehouse of specific knowledge about its cities and climates, its cultural geog-

---

[2]John Friedmann, Regional Development Policy: A Case Study of Venezuela (Cambridge, Mass.: MIT Press, 1966), Part 1; and "A General Theory of Polarized Development," in Niles M. Hansen (ed.), Growth Centers in Regional Economic Development (New York: Free Press, 1972), pp. 82-107; Tormod Hermansen, "Spatial Organization and Economic Development--The Task and Scope of Spatial Planning" (Geneva: UNRISD, 1969), and "Development Poles and Related Theories: A Synoptic Review," in Hansen (op. cit.), pp. 160-203; E. A. J. Johnson, The Organization of Space in Developing Countries (Cambridge, Mass.: Harvard University Press, 1970); Brian J. L. Berry, "City Size and Economic Development," in L. Jakobson and V. Prakash (eds.), Urbanization and National Development (New York: Sage, 1971), pp. 111-143; "Hierarchical Diffusion: The Basis of Developmental Filtering and Spread in a System of Growth Centers," in Hansen (op. cit.), pp. 108-138, and "Relationships between Regional Economic Development and the Urban System: The Case of Chile," Tijdschrift voor Economische en Sociale Geografie, Vol. 60 (1969), pp. 283-307; Paul Ove Pederson, "Innovation Diffusion within and between National Urban Systems," Geographical Analysis, Vol. 2 (1970), pp. 203-254.

raphy and regional divisions. I see no benefit in such a recapitulation for a symposium of this kind. In addition, there will be no attempt to disguise a prescriptive interest. This is not meant to detract from the important individual contributions made by geographers working in foreign areas nor does it imply that all such work should immediately be reoriented in the direction suggested herein. But to sit back complacently and conclude that all is well in the world of regional geography, let's keep doing what we are doing, I find intellectually sterile, dishonestly evasive, and, ultimately, boring.

Consequently, only a restricted portion of the larger geographical literature on Africa will be examined.[3] But it is hoped that by choosing such a specific focus, the contributions which will be of major interest to other area specialists and to the more general theme of geography and area studies will be adequately represented.

Studies of Spatial Development in Africa

One of the most significant developments in the field of African geographical studies over the past decade has been the emergence of a cluster of related research analyzing the patterns of spatial change associated with the processes of social, economic, and political development.[4] These studies have been de-

---

[3] For the most recent bibliographic survey of African geographical research, see the second edition of Sanford H. Bederman, A Bibliographic Aid to the Study of the Geography of Africa, Bureau of Business and Economic Research, Georgia State University, 1972). This volume includes 2, 690 English language citations.

[4] Some representative works in addition to these referenced later in this paper include Akin Mabogunje, Urbanization in Nigeria, University of London (1968), and "Urbanization in Nigeria: A Constraint on Economic Development," Economic Development and Cultural Change, Vol. 13 (1965), pp. 413-438; Peter R. Gould, The Development of the Transportation Pattern in Ghana, Northwestern University Studies in Geography, No. 5 (1960) and "The Structure of Space Preference in Tanzania," Area, Vol. 1 (1969), pp. 29-34; Daniel Ola, "Perception of Geographic Space in Lagos and the Western State of Nigeria," unpublished master's thesis, The Pennsylvania State University, 1969; Michael L. McNulty, "Urban Structure and Development: The Urban System of Ghana," Journal of Developing Areas, Vol. 3 (1969), pp. 159-176; Edward W. Soja, "Communication and Territorial Integration in East Africa: An Introduction to Transaction Flow Analysis," East Lake Geographer, Vol. 4 (1968), pp. 39-57; Richard R. Brand, "The Spatial Organization of Residential Areas in Accra, Ghana, with Particular Reference to Aspects of Modernization," Economic Geography, Vol. 48 (1972), pp. 284-298; Robert H. T. Smith and Alan M. Hay, "A Theory of the Spatial Structure of Internal Trade in Underdeveloped Countries," Geographical Analysis (1969), pp. 121-136.

signed to explore the dynamics of spatial systems in an empirical context, to examine how space is organized and how spatial organization evolves under the impress of contemporary forces of societal change (the latter usually being grouped for convenience under the label of "modernization").

Broadly defined, spatial development refers to a set of processes working within a system of regions to structurally transform the spatial organization of society. This transformation is largely the product of innovation diffusion in which the values, attitudes, and material attributes of "modernity"[5] are spatially disseminated through a population, differentially adopted, and eventually incorporated within a territorially defined social system as a primary basis for social, economic, political, and cultural organization.

Underlying all studies of spatial development is the assumption that the organization of space both affects and reflects the basic processes which operate within society. These processes--migration, urbanization, industrialization, national integration, acculturation--assume a spatial dimension in two important ways. First, they are space-forming in that they work to shape and structure human interaction in space--in the development of transport and communication networks, in the growth of urban and administrative systems, in the evolution of a differentiated and integrated space economy, in the territorial distribution of political authority, in the patterning of cultural traits. At the same time, they are space-contingent. Their space organizing influence is itself shaped by the existing spatial framework. The decision to locate a modern factory, for example, or an administrative center or a cathedral, each a powerful and persistent influence on spatial organization, will depend in large part upon earlier locational decisions and existing patterns of distribution and interaction.[6]

---

[5]Modernity, as it is used here, is a relative term referring to a cluster of associated innovations which, at a given point in time, represents a predominant and expanding paradigm for societal organization. For many millennia before the time of Christ, modernization was directly associated with the diffusion of agricultural knowledge and technology and the organizational and demographic change associated with an agricultural way of life. Twentieth century modernization, in comparison, involves the diffusion of modern science, industrial and communications technology and associated organizational and behavioral changes. To the extent that early agricultural expansion in Europe could be looked at as "Middle Easternization," twentieth century modernization can be viewed as "Westernization."

[6]These terms are taken from John Friedmann, "A General Theory of Polarized Development," *op. cit.*

Origins and Orientations

Spatial development studies have, to a very great extent, sprung from an explicit attempt to apply the models, perspectives, and techniques of modern geographical analysis to Africa. This concern for such phenomena as generalizable patterns of spatial organization, the growth and integration of urban and transport systems, and models of spatial diffusion, did not arise simply from the desire to use the continent as a laboratory for theoretical geography but grew primarily from a genuine belief that modern analytical geography, particularly that portion dealing with spatial or locational analysis, had a great deal to offer to African studies.

As a consequence, spatial development studies assumed a rather distinctive character and orientation. There was, for example, a much greater attempt than in earlier geographical studies of development problems to relate to a wider social science literature, particularly that portion dealing with the modernization process in developing countries. Many studies were specifically designed to draw the attention of political scientists or sociologists to the spatial perspective, to its general neglect in African studies, and to its potential contributions to the interdisciplinary study of societal change.

Another characteristic was a pronounced emphasis on quantification and experimentation with the wide range of statistical techniques used in geography. Given the multivariate nature of the problems being studied, such techniques as principal components analysis and associated methods of quantitative regionalization proved particularly useful. But at one time or another, nearly all existing techniques for the analysis of geographical data were tested in Africa, with varying substantive and technical success.

The concept of modernization was also given a distinctive interpretation. Because of the emphasis on measurable spatial phenomena, there was a tendency to view modernization primarily in infrastructural terms--that is, to assume modernization to be directly associated in space with the multiplication and spread of such external manifestations as post offices, roads, hospitals, schools, administrative offices, commercial activities, and urbanization. This did not reflect some naive notion that the behavioral changes associated with modernization came automatically with the establishment of a post office or when a particular threshold of urbanization was reached. But given the near impossibility of tapping behavioral change directly, especially at the macrospatial level, there was justifiable confidence that societal modernization would be reflected fairly closely in the spatial distribution of those measurable locational phenomena.

Modernization was also viewed, for the most part, as an empirical process without any overt attempt to attach an ideological component or to assume that what was happening was inherently "good." There was no assessment of whether modernization was accompanied by an increasing "demand for self-respect" or the spread of freedom and happiness. This did not imply a denigration of value-explicit approaches to modernization but reflected an expressed desire to focus on spatial change and its effects on such macro-level phenomena as economic development, political integration, urbanization, and nation-building in general.

One can unravel more specific strands of thought which affected the growth and orientation of spatial development studies. A series of articles which appeared in the early 1960's, for example, appear to have been particularly influential. In Taaffe, Morrill, and Gould's "Transport Expansion in Developing Countries,"[7] a stimulating historical-descriptive model of the growth of transport networks was presented based upon empirical research done in Africa and Latin America. Its heuristic unfolding of a process of spatial change permitted the separate detailed histories of road and rail construction in African countries to be fused together in a comparative spatial focus and provided a provocative impulse to discover correlative models of development in a spatial context. From this source, spatial development studies in Africa gained a strong emphasis on structural evolution and, indirectly, a respect for the degree to which the colonial experience could distort the ideal/typical sequences of spatial change.[8]

A second influential article was actually a short note by Peter Gould on the diffusion of development, which appeared in the Journal of Modern African Studies.[9] In this note, Gould called for a greater emphasis on the spatial dimension within the interdisciplinary study of modernization and social change. More specifically, Gould drew attention to the growing literature on spatial diffusion and suggested the potential value of viewing development as a spatial diffusion process. Gould was to preview a persistent emphasis in the spatial development literature on the diffusion theme and a concern for relating geographical studies

---

[7]E. S. Taaffe, R. L. Morrill, and P. R. Gould, "Transport Expansion in Underdeveloped Countries: A Comparative Analysis," Geographical Review, Vol. 53 (1963), pp. 503-529.

[8]See, for example, Edward W. Soja, "Communications and Change," in John N. Paden and Edward W. Soja, The African Experience, Vol. I: Essays (Evanston: Northwestern University Press, 1970), pp. 359-380.

[9]"A Note on Research into the Diffusion of Development," Journal of Modern African Studies, Vol. 2 (1964), pp. 123-125.

of development to the associated literature of other social science disciplines.

Finally, two articles which had their empirical base in East Asia were to have a lasting impact on African geographical studies. Both focused on periodic markets, the first, by Stine, representing an attempt to develop some theoretical notions on the functional bases of market periodicity and its association with central place theory;[10] the second, by Skinner, dealing with much larger questions of rural spatial organization in China and its impact on human behavior and planning.[11] These studies sparked an interest in local level investigations of spatial organization and change which complemented the more macrospatial approaches to development, especially with regard to the structure and growth of urban systems.[12]

It is not possible to review the full range of spatial development studies here, although there is a clear need for such a synoptic and synthesizing exercise. Instead, one subsection of the larger research cluster which has begun to assume a distinct identity of its own, to influence research in other disciplines, and to create critical discussion within geography will be examined in some detail. This is the group of studies of what has been called the "geography of modernization."

---

[10]J. H. Stine, "Temporal Aspects of Tertiary Production Elements in Korea," in F. R. Pitts (ed.), Urban Systems and Economic Development (Eugene: School of Business Administration, University of Oregon, 1962), pp. 68-88.

[11]"Marketing and Social Structure in Rural China," op. cit.

[12]This interest has been particularly focussed on the spatial aspects of periodic markets in Africa. A sampling of the most theory-oriented works on periodic markets would include N. A. Alao, "Periodic Markets in Western Nigeria: Theory and Empirical Evidence," Research Report No. 42, Department of Geography, Northwestern University, 1968; B. W. Hodder and U. I. Ukwu, Markets in West Africa (Ibadan: Ibadan University Press, 1969); Robert H. T. Smith, "The Theory of Periodic Markets: Consumer and Trade Behavior," Preconference Publication of Papers, Canadian Association of Geographers (Waterloo, 1971), and "West African Market-Places: Temporal Periodicity and Locational Spacing," in C. Meillassoux (ed.), The Development of Indigenous Trade and Markets in West Africa (Oxford: Oxford University Press, 1971); Thomas H. Eighmy, "Rural Periodic Markets and the Extension of the Urban System: A Western Nigeria Example," Economic Geography, Vol. 48 (1972), pp. 299-315; R. T. Jackson, "Periodic Markets in Southern Ethiopia," Transactions of the Institute of British Geographers, Vol. 53 (1971), pp. 31-42. For a general review, see R. J. Bromley, "Markets in the Developing Areas: A Review," Geography, Vol. 56 (1971), pp. 124-132.

The Geography of Modernization

More than any other subgroup of spatial development studies, that dealing with the geography of modernization (GOM) has been the direct product of the set of influences and orientations discussed in the preceding section.[13] As such, it exemplifies some of the greatest strengths of the spatial development approach as well as its weaknesses and shortcomings. GOM studies have been remarkably successful, for example, in revealing the rich reservoir of good statistical data available in Africa and illustrating the effective use of such data in the quantitative analysis of spatial change. At the same time, however, many of the studies appear somewhat overburdened with statistical methodology and often seem to suggest to many readers that conceptual advancement can be equated with the introduction of still another statistical technique.

GOM studies have demonstrated the utility of the spatial perspective in the interdisciplinary study of societal change and the relevance of the so-called "new" geography to African studies in general. Yet, until recently, there were few attempts to direct research findings toward more practical and policy-oriented questions of regional development planning. Furthermore, despite an implicit comparative focus, there has been little direct comparison made between the findings of the various studies.

The central focus of the geography of modernization has been to describe the areal incidence of development, generally at the national scale, to explain how this fundamental geographical pattern has evolved over time, and to interpret its implications with regard to various problems of nation-building. The major analytical tool to emerge from this process has been the modernization surface, a simple map of areal variations in levels of development derived from a components analysis of statistical data measuring such variables as urbanization, transport accessibility, and the presence or absence of such indicators of

---

[13] Edward W. Soja, The Geography of Modernization in Kenya (Syracuse: Syracuse University Press, 1968); J. B. Riddell, The Spatial Dynamics of Modernization in Sierra Leone (Evanston: Northwestern University Press, 1970); Peter Gould, "Tanzania 1920-63: The Spatial Impress of the Modernization Process," World Politics, Vol. 22 (1970), pp. 149-170; Thomas H. Eighmy, "Modernization in a Regional Context: Pretheory and Practice in Western Nigeria," unpublished Ph. D. dissertation, The Pennsylvania State University, 1969; Burton O. Witthuhn, "The Spatial Integration of Uganda as a Process of Modernization," unpublished Ph. D. dissertation, The Pennsylvania State University, 1968). A. H. Jones, "Spatial Patterns of Modernization in Uganda," Research Papers, Makerere University, 1971. In preparation is a major study on Tanzania by Anthony DeSouza, Department of Geography, University of Minnesota.

modernization as post offices, airports, administrative centers, hospitals, and schools.

The modernization surface has been a crude tool, severely restricted by data availability, measuring infrastructural reflections of modernization rather than actual behavioral or societal changes. Yet it has consistently provided empirically verifiable insight into the patterns and processes of spatial change and a viable foundation for further analysis. Despite recognized deficiencies in the selection of variables and mode of construction, the modernization surface appears to tap a major component in the spatial organization of society which had hitherto been neglected. To those concerned with the spatial analysis of contemporary Africa, it achieved a status comparable to the population distribution map as the most basic explanatory focus for geographical study.

The ways in which the modernization surface has been used and interpreted provide an interesting developmental sequence in itself. The first modernization surfaces, although not specifically labelled as such, appeared in my study of Kenya in 1968.[14] They represented the first two dimensions of a components analysis and portrayed (a) the general pattern of infrastructural development in the country and (b) the extent of spatial dualism or separatism in development patterns arising from European compartmentalization of the Kenya Highlands. Here the modernization surface was used as a statistical capstone for a more traditional description of social, cultural, economic, and political developments in Kenya. The primary focus of the study was on patterns of national integration and the modernization surfaces were used to construct a composite map, essentially non-quantitative, of the Kenya nation in a spatial context.

The first use of the term modernization surface came in Riddell's work on Sierra Leone,[15] which attempted to assess the "spatial dynamics" of modernization more quantitatively through the use of such techniques as trend surface analysis and a more explicit concentration on diffusion processes. Most of the same variables used to construct the modernization surface were subsequently reevaluated in a multiple regression analysis of migration patterns in Sierra Leone, a subject of continuing interest in Riddell's later research.[16]

---

[14] The Geography of Modernization in Kenya, op. cit.

[15] The Spatial Dynamics of Modernization in Sierra Leone, op. cit.

[16] See J. B. Riddell and Milton E. Harvey, "The Urban System and the Migration Process: An Evaluation of Step-Wise Migration in Sierra Leone," Economic Geography, Vol. 48 (1972), pp. 270-283.

A major step forward in the analysis of modernization surfaces was made by Gould in a study on the spatial impress of modernization in Tanzania.[17] By presenting a series of comparable surfaces over time, Gould was able to illuminate more clearly the major processes underlying spatial change.

> What we have seen emerging in the space called Tanzania between 1920 and 1963, and underlying the very modernization process we have tried to measure and structure, are the dynamics of that most spatial of systems--the central place structure; a system of nodes of various sizes, at differing distances, linked with varying intensities, influencing areas contiguous and between, structuring, focusing and serving their hinterlands, and acting as emergent poles of attraction for the surrounding population . . . And so, underlying any theory of modernization in space and time, there must be a deeper theory of the dynamics of central-place system.[18]

At the same time, however, Gould clearly noted the deficiencies which such an illumination revealed in existing geographical theories applicable to the study of development.

> Such a theory we do not have. Except for a few Monte Carlo simulations of developments in nineteenth-century Europe, our theory is static and geometric, virtually ignoring the concomitant development of transport and communication linkages, the very channels that allow the time-frozen patterns in space to become systems of human interaction and development. But there is increasing doubt that disaggregation of the spatial system into static patterns of nodes, linkages, and movement will lead to the writing of illuminating theory, for every piece of empirical evidence we have confirms the tightly interlocked nature of all these aspects.[19]

In terms of future developments, it has become increasingly clear that it is no longer sufficient to produce evocative descriptions of the historical evolution of modernization surfaces without either grounding these essentially structural pictures in a more appropriate theoretical foundation than has hitherto been used or evaluating them within a more explicitly comparative framework. The metaphor of the modernization surface appears to be well established. The grand panorama of modernization spreading outward from core regions both in a geographically defined diffusion wave and hierarchically through the urban/administrative system seems adequately described. But precisely how and why does this occur? If diffusion is taking place, why is it that development continues to pile up in core regions? Whereas the theoretical literature in loca-

---

[17] "Tanzania 1920-63 . . . ," op. cit.

[18] Ibid., p. 169.

[19] Ibid.

tional analysis and regional science seems to suggest an association between development and the progression toward some hypothesized spatial equilibria, the evidence from Africa and elsewhere suggest almost the opposite. Why? How can we compare the patterns of modernization in Kenya with those in Sierra Leone? What implications do these patterns have for development planners?

One attempt to answer some of these questions by linking geographical studies of modernization more directly with regional development theory has recently been made by Soja and Tobin.[20] Although published in a collection of articles on political development, this study marks a shift away from an emphasis on the political science literature on modernization so influential in earlier GOM studies and towards the work of a group of regional development specialists who have in recent years been attempting to build a spatially-oriented and broadly based theoretical framework for regional development analysis and planning. In particular, it revolves around the contributions of John Friedmann, whose earlier work on Venezuela not only included a qualitative mapping of modernization but also introduced a model of core-periphery relations which has great value in assessing spatial change. The past failure to examine the relevance of Friedmann's work to the geographical study of modernization in Africa represents an excellent example of the communications gap which exists among different area specialists.[21]

Soja and Tobin explore the implications of the core-periphery and related models as a conceptual framework for a more process-oriented and comparative analysis of spatial development. In particular, an emphasis is placed on the tendency, repeatedly verified in the empirical literature, for regional inequalities, primacy in city-size distributions, and other manifestations of spatial disequilibrium to become amplified rather than reduced during the course of regional development, at least until advanced levels of development are approached. Indeed, progressive development concentration in core areas is viewed as a major driving force for change, be it evolutionary or revolutionary. The process is one in which peripheral areas become organized into increasingly core-dependent markets, supply sources, and administrative units while at the same time

---

[20]Edward W. Soja and Richard J. Tobin, "The Geography of Modernization: Paths, Patterns, and Processes of Spatial Change in Developing Countries," in G. Brewer and R. Brunner (eds.), <u>A Policy Approach to the Study of Political Development and Change</u> (New York: Sage Publications, forthcoming, 1973).

[21]Although Latin Americanist geographers have themselves failed until recently to give Friedmann's works the attention they deserve.

there is a limited extension of development impulses from the core, mobilizing the peripheral populations and creating a counter-elite capable of challenging the social bases of integration in the spatial system.

Implicit in this view of regional or spatial development is a recognition of the degree to which such empirical realities as cultural heterogeneity, economic dependence, colonial experience and, in general, the different global context in which development is taking place today in the Third World, necessitates major modifications in Western-based development theory. In geographical terms, it calls for a new kind of grounded theory, devoid of "isolated states," which explores not only the equilibria derived from the assumption of isotropic conditions but more cogently the variable patterns of spatial change arising in specific kinds of cultural and regional contexts.

In essence, what is being suggested is the theoretically-based, practically focused, comparative regional perspective discussed earlier in this paper. Of all the criticisms which can be made of the GOM literature--its tendency to overemphasize urban vs. rural measures of modernization, the inadequacies of the principal components mapping technique, its neglect of more personal or behavioral implications, the failure to include an explicit ideological perspective or greater attention to the more "vertical" relations of man and environment,[22] and what has been perceived by some as overenthusiasm, excessively florid language, and presumed snobbery toward more traditional geographical approaches[23] --the most incisive involves its failure to move beyond technical experimentation and lucid description to associate itself with the larger literature concerned with regional development theory, policy, and planning.

Spatial Planning

A greater urgency has been attached to the need to build a more grounded theory of regional development by the almost explosive growth of interest in what has been called "spatial planning." Spatial planning, most commonly

---

[22] Philip Porter, in a recent review of The Geography of Modernization in Kenya, poses a number of stimulating questions regarding the roles of "geography defined as the analysis of spatial systems and geography defined behaviorally as man-environment interaction" in the study of modernization. See Geographical Analysis, Vol. 5 (1973), pp. 67-73.

[23] See, for example, the oddly ambivalent critique of the geography of modernization literature contained in H. C. Brookfield, "On One Geography, and a Third World," Transactions of the Institute of British Geographers, Vol. 58 (1973), pp. 1-20.

associated with a policy of "growth poles" or "growth centers,"[24] involves an active attempt to manipulate the organization of space to attain specific developmental goals. Spatial organization is looked upon as a powerful and dynamic influence on the development process, not as a passive "given" which requires occasional readjustment to accommodate spontaneous growth or general economic policies.[25]

More specifically, spatial planning usually takes the form of intervention in the urbanization process through the selective concentration of development inputs in certain types of centers. This, it is presumed, will "streamline" the urban system and encourage a more efficient and rapid diffusion of development outward from core areas and downward through the urban hierarchy. If successful, system-wide development will be accelerated, regional inequalities will be significantly reduced, and the distribution of employment opportunities will be brought in closer accord with the distribution of employment seekers.

The growth center approach to spatial planning was originally developed in more economically advanced countries to deal with the persistence of pockets of economic backwardness and the perceived need to deconcentrate development from major core regions. Given the similar, but potentially even more disruptive, problems of core region concentration and peripheral backwardness existing in developing countries, spatial planning in one form or another has diffused rapidly throughout the Third World. As an alternative to the radical reordering of relationships between the developed and developing worlds, which may in the long run prove to be the only means of solving these problems, many countries have chosen to adopt spatial planning policies as a potentially useful <u>internal</u> strategy for development and regional welfare.[26]

---

[24] Growth poles, as originally defined by Francois Perroux, are growing complexes of industries located in abstract economic space and are not necessarily manifested directly in geographic space. Subsequent work, however, has blurred the distinction between a growth pole and its spatial analogy, the growth center, to the point that both are often used interchangeably when referring to development in geographic space. For some of the methodological problems involved here, see Tormod Hermansen, "Development Pole and Related Theories: A Synoptic Review," <u>op. cit.</u>, and D. F. Darwent, "Growth Poles and Growth Centers in Regional Planning--A Review," <u>Environment and Planning</u>, Vol. 1 (1969), pp. 5-31.

[25] For a detailed discussion of the concept of spatial planning, see Tormod Hermansen, "Spatial Organization and Economic Development: The Task and Scope of Spatial Planning," <u>op. cit.</u>

[26] One of the earliest attempts at spatial planning in Africa is found in

Spatial planning takes on an additional, although related, role in Third World countries. In Africa, for example, there is a particular need to restructure the spatial legacy of colonialism, which has so powerfully shaped the geography of modernization, to suit the requirements of independent, African-governed countries.[27] Little attempt was made during the colonial period to promote the kinds of regional and cultural integration necessary to support stable, independent states. The colonial space economy was oriented to external markets, transport systems were characteristically tree-like in structure, draining and funneling resources rather than facilitating interregional contacts. The distribution of social services, indeed the whole pattern of infrastructural modernization, was geographically biased to reflect initial colonial locational decisions, administrative convenience, and the demands of exploitation and pacification. Whether this pattern of spatial "investment" can be significantly modified without curtailing national economic development rates is still a moot question. But it exists as a major problem for spatial planning, especially when the reduction of regional development inequalities assumes greater priority in national planning strategies.

What is of most concern for our present purposes is not that spatial planning is misguided or incorrect in its general assumptions and objectives. There is sufficient support, both empirical and theoretical, to accept the kinds of broad relationships presumed to exist between the structure of the urban system and the spatial diffusion of development. There seems little doubt, in my mind at least, that some policy of selective concentration (or deconcentration) can work effectively toward speeding up the spread of development and reducing regional imbalances if not completely blocked by inequities of the international economic system.

But what are the specific policies to be followed? Take, for example, the question of primate cities, one of the most controversial subjects in spatial

---

David Grove and Laszlo Huszar, The Towns of Ghana, The Role of Service Centers in Regional Planning (Ghana University Press, 1964). Currently, the Government of Kenya has embarked on an ambitious program of regional development based on some simplified concepts of central place theory and has published a series of provincial physical development plans detailing the present and proposed functional structure of the Kenya urban system. See Government of Kenya, Development Plan 1970-74, and the various provincial plans, published by the Ministry of Lands and Settlement.

[27] M. I. Logan, "The Spatial System and Planning Strategies in Developing Countries," Geographical Review, Vol. 62 (1972), pp. 229-244.

planning. Is urban primacy in itself detrimental to development, thereby necessitating large-scale programs to disperse new industry, control migration, or establish massive new alternative growth foci? Or is primacy an efficient and perhaps essential short-term stimulus to development, attracting innovative activities and promoting growth which under other conditions could not be effectively supported--at least until a sufficiently high level of development is reached to reduce regional inequalities more efficiently, economically, and "naturally"? Or will the answer depend entirely on such variables as cultural homogeneity, resource endowment and location, and the territorial distribution of political power?

Similar questions can be asked of other specific policies. Is it more appropriate to focus on smaller centers, reorganizing the rural landscape to create a more concentrated clustering of development and thus reduce massive migration to the major metropoli--the "centralizing upward" approach of E. A. J. Johnson?[28] Or is it better to "decentralize downward," as suggested by other planners,[29] by encouraging the movement of larger scale industry from the biggest cities to the next largest order? Or is it best to focus on intermediate size centers to create the necessary transmission points for the diffusion of development from core to periphery?[30] We still do not know the ways in which growth centers affect their surrounding region, whether certain kinds of transport improvement accentuate or reduce core region concentration and the intensity and areal extent of "backwash" effects, or if the emergence of a regional growth center will significantly absorb or further encourage migration to overburdened primate cities.

None of these questions can be properly answered without the growth of a more empirically grounded theory of regional or spatial development and such a theory will not emerge without a succession of detailed comparative studies exploring, with an explicit conceptual focus, the historical paths, patterns, and

---

[28]E. A. J. Johnson, <u>The Organization of Space . . .</u>, <u>op. cit.</u>, and <u>Market Towns and Spatial Development in India</u> (New Delhi: National Center for Applied Economic Research, 1965).

[29]For example, John P. Lewis, "The Problem of Growth Centers," paper read to the Hyderabad Seminar on Accelerating District Industrialization, 1966, referred to in Berry, "Relationships between Regional Economic Development and the Urban System," <u>op. cit.</u>

[30]Niles M. Hansen, <u>Intermediate-Size Cities as Growth Centers</u> (New York: Praeger, 1971).

processes of spatial change in specific areas of the world. Furthermore, the appropriateness of any given policy cannot be gauged solely in terms of ceterus paribus abstractions from reality but must be evaluated within particular cultural, political, economic, and social contexts. Therein lies the primary goal of comparative regional analysis.

A Concluding Challenge

No other phenomenon of recent origin is so replete with implications for foreign area studies in geography (or regional geography in general) as the growth of an explicit spatial emphasis in the regional development literature. The initial impetus for this orientation was provided by the perceptive but limited excursions into the spatial domain made by such development theorists as Alfred Hirschman, Gunnar Myrdal, and Francois Perroux in the 1950's. Today, however, the full breadth of geographical theory and methodology is being sensitively and selectively explored to the point that major contributions to the synthesis, explication, and application of geographical concepts are being made by scholars who do not consider themselves primarily as geographers.

It is indeed ironic that the first book to deal expressly with the organization of space in developing countries was written by a non-geographer.[31] A simple perusal of the table of contents of this book makes the irony even more pointed, especially to the strongly spatially-oriented geographer. There are chapters and sections dealing with the nature of landscapes in human geography, market systems and spatial design, central place theory and the regional organization of space, spatial reconstruction, spatial experiments, and directing creative talents to spatial problems. There are some weaknesses in the book. Large portions of relevant geographical literature (particularly that dealing with diffusion processes) appear to be overlooked and one senses a somewhat limited grasp of the full implications of the spatial perspective. Yet the book is thoughtful, provocative, and geographical to its core.

It is not coincidental that a presentation aimed at exploring the contributions of geographers to area studies and the impact of African geographical research on geography as a whole should conclude with a brief tribute to a work done by an economic historian and specialist on South Asia. It illustrates a

---

[31] E. A. J. Johnson, The Organization of Space . . . , op. cit. Irony is similarly noted in Peter Gould's review of this book in Economic Geography, Vol. 48 (1972), p. 444.

strong personal feeling that the major challenge facing area studies in geography today lies in blending a deep and detailed knowledge of a particular cultural realm with an awareness of and ability to build upon the conceptual and theoretical foundations of modern analytical geography, and to direct this blend more effectively toward the practical and theoretical problems of comparative regional development.

CHAPTER 9

THE CONTRIBUTION OF GEOGRAPHY TO FOREIGN AREA STUDIES: THE CASE OF SOUTHEAST ASIA

C. A. Fisher
School of Oriental and African Studies, London

A British contribution to the assessment of the part played by geography in foreign area studies may present a rather different viewpoint from that of Americans, if for no other reason than that area studies were devised in the United States and were for many years regarded with suspicion and even open hostility in British and other European academic circles. However, I wish to state at the outset that I personally never shared these attitudes and indeed have throughout been a consistent supporter and practitioner of the area studies approach, which provides an ideal intellectual setting in which to pursue my own geographical interests in Southeast Asia.

The Origin and Rationale of Area Studies

The modern concept of area studies developed during the Second World War out of the American realization that the United States would need to concern itself much more with the world outside its borders after the war was over. To do this effectively Americans would therefore need as a matter of urgency to expand the as yet relatively limited scale of their expertise on the non-Western world, and especially on those parts of it which seemed most likely to be "problem areas." Thus area studies came into being as the means of implementing a crash programme to supply this need, and although its earliest emphasis was on the areas presenting the most immediate problems, namely the Asian-Pacific theatres of war and the Soviet Union, the rapidity and extent with which social, economic, and political problems began to proliferate after 1945 soon led to the extension of area studies programmes to cover the whole of the non-Western world and even marginal parts of the West itself.[1] After an interval of several

[1]The U.S.S.R. is such a case, as also is Latin America insofar as the

years similar area studies programmes began to be developed in some British and continental European universities and also in Canada, Australia, and New Zealand, so that for a time in the 1950's area studies might have been defined as the study by the West of the non-Western world. But more recently Japan, India, Singapore, Ghana, and other Asian and African countries have breached the West's monopoly in this respect, and this development, which is already contributing new and valuable insights, should still further enrich our work in the future.

Nevertheless the area studies concept was basically American and it grew out of a frank recognition of America's own unfamiliarity with the areas in question. This at least was refreshingly modest, and much to be preferred to the fractious and pretentious claims of some European[2] scholars to virtual omniscience in respect of areas over which their own respective countries had hitherto ruled. For even though the European imperial countries did unquestionably possess unequalled knowledge in many important respects, this was all too often slanted by conscious or unconscious bias in favour of official policies, and so rigidly confined within the arbitrary pigeon holes produced by colonial boundaries as to be dangerously out of focus in the Asian or African context. And while Americans also had their prejudices, arising most obviously from an inbuilt anti-colonialism and a related tendency to underrate the obstacles to innovation in societies both different from and in many cases much older than their own, these views were so explicit that appropriate allowance could be made for them. In any case Americans were willing to learn and even to import many of the less old-fashioned European experts to teach them. On balance, therefore, the American initiative in promoting area studies was thoroughly beneficial, and indeed came as a great gust of fresh air to many dusty academic corridors. A new look, particularly at oriental studies, was long overdue.

The most fundamental question posed by the new area studies is implicit in its name: on what basis are the areas in question to be defined? Or, to adapt a concept of Toynbee's, what kind of area constitutes an intelligible field of study within the non-Western World?

---

emphasis is placed on its Spanish and Portuguese rather than its indigenous heritage. In the area studies context it is arguable that the purely geographical term Central and South America would be better than the implicitly cultural Latin America.

[2]The term European of course includes the British both before and after they were dragged kicking and screaming into the Common Market, on 1 January 1973.

The orthodox approach to such studies in Europe before the Second World War had been modelled on the traditional academic way of studying Europe itself. Until the 19th century this had been almost exclusively that of the "humanities," a term which, though devised to signal a break-away from the earlier totalitarian control of divinity, nevertheless tended to imply that only Europeans were human, and indeed that the only human beings who really mattered were those of classical Greece and Rome.

The great merit of this classical humanities approach was that it dealt with something that had long since run its course and was therefore complete. In the perhaps not strictly immortal words of Edmund Clerihew Bentley[3]

> There's a lot to be said
> For being dead.

Being thus complete and neatly circumscribed in both place and time, the classical civilizations of the Mediterranean world could be studied in their totality, in depth, and with detachment. To do this called first for a knowledge of their languages, Greek and Latin, which, being dead, were also from the pedagogue's point of view perfect, since he had the last word on them and so could systematize them to a degree never attained by anyone who had ever spoken either of them as his native tongue. But besides their resultant value as a precise discipline, Greek and Latin also unlocked the door to the minds of the supposedly superior classical thinkers. And since the classical world was essentially an unspecialized world, its thinkers ranging uninhibitedly over the whole field of thought, including mathematics, philosophy, art, literature, politics, and law, the study of the classical humanities was believed to provide the perfect education for a gentleman, whether of leisure or involved in public affairs.

As a result of the profound changes in everyday life produced by the Industrial Revolution, the classics began in the late 19th and early 20th centuries to be replaced as the basis of a general education by other somewhat more up-to-date subjects, in particular post classical--or in other words medieval and modern--history. But this in fact retained much of the spirit of the classics in that it was essentially a study rooted in language training and concerning itself almost exclusively with European man, though the latter was no longer confined to the Greco-Roman world, and had even extended his activities beyond the boundaries of Europe itself. But Western Europe remained the focus, and even

---

[3] However, these remarks were made in an Asian context, being in fact preceded by "What I like about Clive/Is that he's no longer alive."

in planning the Cambridge Modern History, which purported to be "not only modern but universal," Lord Acton stated in 1907 that it "should include no Russian chapter until Russia becomes a factor in general history under Peter the Great."[4] Thus, as Professor Arthur J. Taylor aptly commented, "With such terms of reference Chinese history could safely be left to the Sinologist and Africa even more firmly consigned to the anthropologist."[5]

This indeed was precisely what had been happening. Over a period of centuries a handful of European scholars had gradually developed Oriental Studies on the pattern of the classical humanities. The theologians' need to learn Hebrew as a key to the Old Testament as Greek was to the New; the great missionary enterprise, particularly of the Jesuits, in Asia, culminating in their attempt to infiltrate Christianity into China by rigorously training missionaries in the Confusian classics so as to be able to talk on equal terms with the Chinese *literati*; and, beginning in the 19th century, the employment in colonial administration and diplomatic service in various parts of Asia of many well educated--i.e., classically trained--men whose study of the local vernaculars led them on to such ancient languages as Sanskrit, Pali, and classical Chinese--all played their part in moulding the tradition of such orientalism as existed in a few European--and still fewer American--universities before the Second World War. But no comparable development had taken place in respect of African studies, since Africa was assumed to possess no cultural traditions worthy of investigation by such means. Africa, therefore, together with other parts of the world at a similar cultural level, was left to the new genre of specialists known as anthropologists. And although the literal designation of their field seemed to be virtually synonymous with that of the humanities (except that it derived from Greek, not Latin, a difference which in other contexts was usually regarded as a sign of superiority) it came in practice to mean not the study of mankind in general, but only of those members of the human family who, at least until very recently, have usually been referred to as backward or primitive.

Against this background it is not surprising that even the initial minority of traditional orientalists who evinced sympathetic interest in the new area studies saw the latter merely as an updating of the older classical-oriental

---

[4] Lord Acton, *The Cambridge Modern History: An Account of Its Origin, Authorship, and Production* (Cambridge, 1907), p. 22. Cited in Arthur J. Taylor, *History in an Age of Growth* (Leeds, 1964), pp. 13-14.

[5] *Ibid.*, p. 14.

studies, which would take account of the insights of the social sciences, such as economics, politics, international relations, anthropology, sociology, demography, and perhaps even geography, but would remain fundamentally concerned with the study of one or other of the so-called "great cultures." This view is clearly apparent in the following comments of the late Sir Hamilton Gibb,[6] a distinguished former Professor of Arabic, first in London and then in Oxford, who subsequently became Director of the Center of Middle Eastern Studies at Harvard:

> The important thing is that, if any such synthesis is attempted, it is significant and valuable only if the various data are related to a central core. The Orientalist's function is to furnish that core out of his knowledge and understanding of the invisibles--the values, attitudes and mental processes characteristic of the 'great culture' that underlie the application even today of the social and economic data--to explain the why, rather than the what and the how, and this precisely because he is or should be able to see the data not simply as isolated facts, explicable in and by themselves, but in the broad context and long perspective of cultural habit and tradition.

Clearly, however, if the intelligible field of area studies is to be regarded as coterminous with the area occupied by a single "great culture," not merely must most of Africa be excluded, but so also must an area like Southeast Asia in which several major cultures overlap and interpenetrate. Nor is this merely a matter of academic disputation, as can be seen from the ridicule which Richard Harris of the _Times_ has poured upon what he calls the "geographical fiction" of Southeast Asia. Thus for example,[7]

> What in fact has the West done in face of its past geographical misconceptions and the unhappy legacy of French imperial sprawl? It began by inventing a new and confusing term to replace the valid old one. In place of Indo-China which did, at least, suggest the meeting point of two separate civilizations, we are now saddled with South-East Asia, inherited from a wartime command. . . . But the label South-East divorces the area from both south Asia and east Asia, giving it an identity, and by inference a unity of its own which it has never had. . . .

And a leading article in the _Times_ a year later[8] adopted a similarly muddled argument:

> . . . The danger is that geography may obscure or distort the political, religious, and cultural realities out of which the world's conflicts are born. . . . Why should we let ourselves be governed by cartographical divisions when the divisions of language, religion, and race; of ways of thinking, and ways of

---

[6] Sir Hamilton Gibb, _Area Studies Reconsidered_ (London, 1963), p. 15.

[7] Richard Harris, _Independence and After_ (London, 1964), pp. 9-10.

[8] "In the Wrong Bundle," _The Times_, London, February 22, 1965.

> governing; are so much more potent and lasting, and can form such prodigious dykes that atlases never mark? The world will not settle down peacefully in accordance with our maps. Southeast Asia, in any case, is an upstart cartographically. Before the war the labels were 'Farther India,' 'East Indies,' 'Malay Archipelago,' and, with more approximation to reality, 'Indo-China' (of which French Indo-China, properly so-called, was a part). Alas, just as the Near East went down to the conquering Middle East, so a wartime command has conquered South-east Asia. We might get a little nearer to solving its problems if we looked at the realities behind the geography.

These two statements, which are by no means untypical of the kind of criticism to which area studies have been subjected in Britain, are in fact both ill argued and ill informed. To begin with, geography is not synonymous with cartography, and it is wholly incorrect to suggest that the name Southeast Asia, let alone the concept to which it refers, derives from the setting up of South East Asia Command in 1943. Long before Europeans began to frequent Southeast Asia, this area (which of course includes the Malaysian, Indonesian, and Philippine archipelagos as well as the Indochinese peninsula in the broadest sense of that term) was recognized as such by both the Chinese and the Japanese who respectively called it the _Nan Yang_ and the _Nanyo_, both meaning the southern seas. In modern times the name appeared, with its present-day connotation, in a British Parliamentary Paper, _Trade and Shipping in South-East Asia_, in 1900, and although such authors as Alleyne Ireland and J. S. Furnivall wrote about what they called the _Far Eastern Tropics_ and the _Tropical Far East_, both referred explicitly to Southeast Asia which they recognized as an entity in its own right. And as _Südostasien_ the term was coming into general use among German and Austrian scholars during the inter-war years.[9]

Still more serious, however, in relation to our consideration of area studies, which we have seen tended to be particularly concerned with problem areas, is the glib _ex cathedra_ pronouncement that the world's conflicts are born out of "political, religious, and cultural realities" which, it is claimed, are obscured by geography. Thus, for example, while it is merely tautological to argue that political conflict--which may or may not degenerate into armed conflict--is related to politics, neither religion nor culture[10] as such were of more

---

[9] See, e.g., Robert Heine-Geldern, "Weltbild und Bauform in Südostasien," _Wiener Beiträge zur Kunst und Kulturgeschichte Asiens_, Vol. 4 (1930), pp. 28-78, and Karl J. Pelzer, _Die Arbeiterwanderungen in Südostasien_ (Hamburg, 1935).

[10] In the somewhat archaic sense in which the _Times_ uses this unnecessarily dual terminology.

than marginal significance as causes of the Second World War. Admittedly, the Germans talked a great deal about Kultur (hardly, however, the same thing as culture) but they talked even more in the 1930's about Herrenvolk and Lebensraum, which respectively were racialist and territorial concepts. And essentially similar considerations were adduced by the Japanese in justification of their long sequence of aggression from the war against China in 1894-95, via the 21 Demands of 1915 and the invasion of Manchuria in 1931 to the "Greater East Asia War" against both China and the Western colonial territories of Southeast Asia in 1937-45. And finally, even if for the sake of argument one assumed that culture in the fuller sense was the primary source of conflict, this would surely make it all the more important to study the main geographical areas of cultural overlap, such as the Balkans, in which the First World War began, and Southeast Asia which was the scene of Japan's involvement in the Second.

In fact, however, the supposed antithesis between culture and geography is entirely false. Culture and a fortiori civilization do not descend upon man like manna--or even pennies--from heaven, but are created by particular communities of men in particular geographical and historical contexts. Thus as Hu Shih, one of the leaders of the 20th Chinese cultural renaissance, has observed, "the civilization of a race is simply the sum total of its achievements in adjusting itself to its environment." It is on these grounds that I would argue that in any programme of area studies geography should take precedence over culture. I mean this in the literal sense of coming first, without implying that geography is either more or less significant than what follows, though on purely logical grounds it would seem appropriate, in trying to delineate and study any area, to begin by considering the contribution of a subject whose field has been most consistently defined as the study of places. And here it is appropriate to quote some comments of Jacques Barzun:[11]

> Yet there was one valuable discovery made in area instruction that no one would now dispute about. The solidest thing about an area is its geography, and our people is singularly ignorant of any but its own . . .
>
> . . . It is not a subject for infants only, since everything that happens happens in space. Military strategy, industry, commerce, communication, political feeling, art, and science have locus and momentum. To judge and take part in them, the channels that link them through three elements must

---

[11] Jacques Barzun, Teacher in America (New York, 1947), p. 145, quoted in Edward L. Ullman, "Human Geography and Area Research," Annals of the Association of American Geographers, Vol. 43 (1953), pp. 54-66.

be known. And from the earliest days spent in school it would help reduce that disembodied abstract feeling about subject matter if its details could be fastened down to their point of origin.

From the foregoing discussion it will be apparent that I regard Southeast Asia as providing a test case in respect both of the validity of the area studies approach and also of the contribution which geography can make of it. For while I totally reject the view that Southeast Asia is a "geographical fiction" derived solely from the accidents of war, there is no denying that, alone among all the areas with which this seminar is concerned, Southeast Asia was not generally recognized in the West as a meaningful entity, concerning which it was possible to make useful "middle-order generalizations," until after the Japanese entry into the Second World War. For this reason, therefore, I propose to try to illustrate the kind of insights which I believe the geographer can contribute to area studies by giving a brief outline of the way in which I as a geographer have come to see Southeast Asia.

One Geographer's View of Southeast Asia

The most fundamental fact about Southeast Asia is the extent to which throughout its history it has been overshadowed by its more populous and culturally more advanced neighbours in South and East Asia. These last are easily defined because in each case one great country, by virtue of its vast population and the related depth and continuity of its civilization, has set its own distinctive imprint upon the region of which it forms the central dominant component. By contrast, Southeast Asia is a severely fragmented area, with no great central riverine lowlands such as those of the Indus-Ganga and the Hwangho-Yangtze which have provided the core areas of the respective great cultures of India and China.

Instead, insofar as Southeast Asia has a central component, it consists not of land but of sea, though if the sea level of Monsoon Asia were to fall 100 fathoms as it did in the Quaternary Ice Age this fundamental physical difference between Southeast Asia and its two neighbours would disappear. For while the land areas of India and China would increase by a mere 5-10 per cent, that of Southeast Asia would double, and perhaps more significantly the extent of its lowland area would quadruple as the great expanses of the Sunda Shelf underlying the South China Sea, the Java Sea, and the Straits of Malacca once again became dry land. Thus Southeast Asia would consist of a single great sub continental landmass, comparable in scale to India or China proper, together with a few large offshore islands formed by the coalescing of parts of the Philippines and eastern Indonesia, which together would occupy an analogous position to that of Japan and Taiwan relative to subcontinental China.

In effect, therefore, Southeast Asia is a subcontinent which in geologically recent times has lost its principal lowlands, so that what remains consists preponderantly of rugged uplands. Thus even its largest rivers are merely the former upstream stretches of a series of much longer rivers, whose lower courses once flowed through wide alluvial plains developed on what is now the Sunda Shelf. This explains why the valleys of today's rivers are disproportionately narrow, and mutually isolated from one another by wide expanses of plateaux and ranges, and indeed many of the largest rivers, including the Salween, Mekong, and Songkoi, flow through predominantly gorge-like valleys for four-fifths or more of their course before suddenly terminating in hugh deltas, which have formed extremely rapidly in adjustment to the post-glacial rise in sea level and contain large expanses subject to tidal innundation and/or severe seasonal flooding following the peak period of the monsoonal rains.

While these characteristics are most strikingly exemplified by the major mainland rivers, they are repeated on a smaller scale in the western Indonesian islands. For there also the main rivers flow towards the Sunda Shelf, where they terminate in a series of coastal deltas which, especially in Sumatra and Kalimantan, have merged to form wide belts of swamp stretching continuously for hundreds of miles along the coast.

Thus a high proportion of what appears on the modern atlas map to be the main lowland areas of Southeast Asia have throughout the greater part of its history suffered from acute waterlogging. This in turn is related to the regional climate which again differentiates the region markedly from both of its great neighbours. For whereas practically the whole of China and a large part of the Indian subcontinent (including the entire Indus-Ganga lowlands apart from the coastal half of the Ganga delta) lie outside the tropics, Southeast Asia straddles the equator and is almost entirely comprised within the humid tropics which, moreover, are exceptionally humid here owing to the extent to which the land surface is interpenetrated by the sea. Except, therefore, for the higher uplands and a few relatively small inland areas of at least seasonally drier climate, the prevailing natural vegetation of Southeast Asia has everywhere been much denser than that of China and of all but a few peripheral parts of the Indian subcontinent.

In short, those wholly untypical parts of Southeast Asia which were not initially extremely difficult to clear and otherwise improve for agricultural use, and at the same time possessed soils of average to high fertility (mainly of alluvial or neutral-basic volcanic derivation) and a reliable seasonal alternation of adequately wet and dry periods for the growth and harvesting of good crops of rice or other grains, amounted in total to only about 5 per cent of the region's overall area. And this very small total was made up of widely scattered patches, primarily in the middle reaches of the Irrawaddy and Chao Phrya valleys, the Cambodian lake basin, the valleys of east-central Java, and the central lowlands of Luzon. To these should also be added the exceptional delta of the Songkoi, which in both climate and terrain has more in common with sub-tropical China than with the rest of Southeast Asia, from which northern Vietnam is separated near the 18th parallel by the Annam Chain, in essentially the same way as the Arakan Yoma separate the Indian sub-continent from Burma in the vicinity of the 21st parallel.

This last point raises the further question of overland access into Southeast Asia, which also presents a striking contrast with conditions in India and China. While both these last were repeatedly invaded by peoples coming immediately from Inner Asia via the gaps in their respective northwest frontier zones, Southeast Asia was shut off from the Inner Asian centres of dispersal by the greatest mountain barrier in the world. Thus it was subjected only to a much more attenuated, though nevertheless prolonged, series of drifts of peoples from the Sino-Tibetan borderlands, who were gradually displaced southwards down its mainland valleys under pressure from the slowly filling demographic reservoir of what today is China proper.

However, while as has already been implied, direct overland movement from India into Southeast Asia is very difficult, access by sea is easy, and although this situation has not led to any mass migration, as was once erroneously believed, it did undoubtedly facilitate the dissemination of a succession of cultures, in the widest sense of that term, eastwards across the Bay of Bengal and along the ramifying seaways of Southeast Asia beyond.

It is against the background of such an environment, humid tropical, heavily forested, and territorially fragmented into a series of lesser compartments draining out to the South China Sea and its various maritime annexes, that one must try to envisage how repeated southward drifts of peoples, initially very primitive but later at successively high levels of culture, began to establish themselves and to eke out their livelihoods in what to them were the more favourable ecological niches available. And during the past two millennia this process was further complicated by the ever-increasing movement of foreign traders, missionaries, and others, coming first from India and later from still remoter areas, along the arterial seaways through the geographical heart of Southeast Asia.

If one accepts Carl Sauer's brilliant hypothesis,[12] the first major advance in this process came when Mesolithic peoples, living in the valleys and coastlands fringing the Southeast Asian seas, initiated mankind's first form of agriculture as part of a culture in which the men made boats and fished while the women raised small livestock and planted yams, aroids, and bananas. Whatever may be the truth in this respect, however, it seems clear that the degree of human control over the Southeast Asian environment remained extremely restricted for a very long time, though a significant advance had apparently been made before the dawn of the Christian era, by which time the inhabitants of these same maritime areas had developed great skill in seamanship and had begun to practice wet rice cultivation. These activities provided the material basis of a much more elaborate culture, the so-called Austro-Asiatic civilization of Georges Coedès, which in varying degree appears to have been ancestral to the traditional way of life of the great majority of present-day Southeast Asians.

Nevertheless this proto-civilization was less advanced than the then contemporary civilization of India and China, and it was not until regular maritime contact was established with India in the early centuries A. D. that Southeast Asian societies became capable of supporting an urban superstructure and began to be reorganized in larger and more effective political units. These earliest Indianized states were originally founded at key points along the sea routes which Indian pioneers were exploring through Southeast Asia to China and, since many of these points were at or near the mouths of important rivers, some early attempts were made to bring parts of the deltas, notably that of the Mekong, under control by means of Indian-style hydro-technology. In the end, however, nature seems to have regained the upper hand and more permanent results were only achieved when Indianized civilization penetrated upstream into areas where the floods were less difficult to control. The most striking example of this was provided by the great Khmer civilization of Angkor in the Cambodian lake basin, though smaller irrigation systems were successfully developed, notably in central Burma and Siam, and also in east-central Java and elsewhere in Indonesia.

Each of these developments played a crucial part in the formation, during

---

[12] Agricultural Origins and Dispersals (New York, 1952).

the period between A. D. 500 and 1500, of one or other of the proto-national states of Southeast Asia, all of which originated well inland where incoming Indianized civilization gradually fused with the older indigenous traditions, though in the two mainland cases the process was complicated by the arrival from the north of new immigrant peoples--Burmese and the Thai--who had already acquired some knowledge of Chinese culture and technology. Meanwhile the Songkoi delta had been brought under effective control by the superior hydrotechnology of the Chinese themselves who had incorporated this northeastern corner of Southeast Asia into their imperial domains in 181 B. C. But in A. D. 939 it broke away and thereafter, as its population gradually began to outgrow this narrowly limited area, the Vietnamese slowly but steadily expanded their territory southwards, eliminating the Indianized coastal Kingdom of Champa, and eventually in the 18th century wresting the Mekong delta from Cambodia, the much diminished successor of the Khmer empire of Angkor, which had been reduced to a shadow of its former self by the Thai invasions of the early 15th century.

The experience of Angkor was in fact merely an extreme case of a more general condition which goes far to explain why, even at the beginning of the 19th century, the population of all Southeast Asia probably did not exceed 25 million, which implied a density of about one-sixth that of India and one-twelfth that of China proper. Fundamentally this arose from the inability of the widely scattered series of small nuclei of population representing a mere handful of clearings in the overall expanse of jungle to gain a lasting measure of control over their environment. With the possible exception of northern Vietnam and of Angkor at its zenith, it is unlikely that any of the five or six major cores of population already referred to ever contained as many as four million people before 1800, and indeed the whole of the Philippine archipelago, the most isolated area of all, which had accordingly been by-passed by Indianization, seems to have contained under two million at that date.

In the still essentially untamed humid tropical environment of pre-19th century Southeast Asia, the occasional exceptionally severe flood, sudden epidemic, or surprise attack by a rival power could have a disastrous effect on such small nuclei of people, and if, by a not unknown chain reaction, all three kinds of disaster should follow swiftly on the heels of one another, the sequence might prove completely catastrophic. In effect, therefore, such conditions produced their own vicious circle. None of the kingdoms based on such small numbers of people was really secure; mutual isolation fostered mutual distrust which was compounded by memories of past rivalries; and so the whole area remained divided and weak, and plagued by intermittent warfare between one state and another. Such conflicts, however, had little or nothing to do with religious or other cultural differences. Indeed, even when a major new cleavage appeared after the 13th-16th centuries, during which Islam had spread over virtually all the southern half of the region while the areas to the north remained Buddhist,[13] local warfare within Southeast Asia was far more common between states of the same religion than between those of different faiths, if for no other reason than that, except along the narrow isthmus of Kra, the wide and often rough expanses of the South China Sea kept the Buddhist and Muslim halves of Southeast Asia well apart from one another.

These same geographical circumstances moreover help to explain why, although the Malaysian-Indonesian archipelagos had seen the rise and fall of several significant thalassocracies, no indigenous attempt was ever made to unite all the lands around the South China Sea, as Rome had once united the Mediterranean basin. For in relation to such a task all the main Southeast Asian centres of population were not merely much too small but also too remote from the critical strategic focus, namely the maritime crossroads where the through sea route via the Malacca Straits and the South China Sea dips south round the tip of the Malay peninsula and so meets the secondary shipping routes through the Sunda Straits and the Java Sea.

In particular, the extremely low population density of pre-19th century Malaya, resulting primarily from the more acute environmental problems presented by truly equatorial as opposed to tropical conditions, has constituted a major geopolitical weakness for Southeast Asia as a whole. By analogy with Preston James's concept of the "hollow frontier" in Brazil,[14] one might thus argue that Southeast Asia's crucial problem was that it possessed a hollow focus, which however is not quite the same thing as a soft centre.

The fact that no Southeast Asian state ever attempted to extend more widely the kind of influence which Sri Vijaya had wielded over the maritime crossroads between the 7th and 13th centuries was of course also related to the activities of other and stronger externally-based powers, including China which in the early 15th century sent a series of naval expeditions to the Nan Yang. But in the event

---

[13]Though most of their peoples changed from the Mahayana to the Theravada School of Buddhism.

[14]Latin America, rev. ed. (New York, 1950), p. 7.

it was the Europeans who, from the time of the Portuguese capture of Malacca in 1511 to the Japanese defeat of Singapore in 1942, came to play the dominant role in this respect.

For the first three centuries this role was essentially negative so far as Southeast Asia itself was concerned. Portuguese, Spaniards, Dutch, and in a lesser degree British and French projected their own rivalries into Southeast Asia and in so doing fomented new quarrels among the local peoples themselves. And while the Dutch, having virtually displaced the Portuguese in the 17th century, developed from their headquarters in Batavia a wide-ranging network of maritime trade routes interconnecting the entire region, this served not to draw the indigenous Southeast Asian peoples together but rather to keep them apart. Only in the hitherto retarded Philippines did the dissemination of Christianity cause any radical changes in the character of society and the introduction from Latin America of maize, cassava, and peanuts significantly diversify the subsistence farming pattern. For neither the mercantilism cloaked in Christianity as introduced by the early Europeans, nor Islam which had preceded it, was accompanied by any fundamental technological changes comparable in ecological significance with those which had marked the great age of Indianization. But more momentous changes were in the making, and during the final century and a half of colonial rule Southeast Asia underwent a far more fundamental transformation than it had ever experienced before.

On the eve of the Japanese invasion at the end of 1941 the population of Southeast Asia stood at 150 million,[15] which represented a six-fold increase over that at the beginning of the 19th century. This phenomenally rapid growth provided unmistakable evidence of the extent to which the region had at last broken through the environmental constraints which had held it back for so long. Basically this breakthrough had been achieved by applying the new Western technology of the Industrial Revolution to the problems of Southeast Asia. This technology had come in primarily via the new three-port-axis--the series of free ports established by the British on the offshore islands of Penang, Singapore, and Hong Kong--which served as both trans-shipment points between the arterial seaway and regional shipping routes and as local centres of equally essential banking and commercial expertise for Southeast Asia. Both the coming of the steamship and the opening in 1869 of the Suez canal greatly assisted the diffusion

---

[15]Of whom approximately 10 million were of Chinese and 3 million of Indian extraction.

of these key Western innovations, and the fact that so high a proportion of the fragmented territories of Southeast Asia lay within easy access of the sea--and indeed of one of the greatest international sea routes in the world--was of immeasurable importance in promoting the economic development (in the old sense of that term) of Southeast Asia as a whole.

Hand in hand with this process came the extension of European rule, spreading inwards from new port cities on or near the coasts, and consolidated by the creation of rail and road systems ramifying into the interior from these same cities which usually functioned as the administrative capitals. With European control came the introduction of law and order, and Western technology was harnessed to bring the great deltas and other hitherto neglected areas into cultivation and to exploit major resources of already known and newly discovered minerals. Meanwhile the stamping out of internecine warfare, the creation of vast new rice bowls, the ability rapidly to move food and other relief supplies to areas hit by crop failure or other calamities, and the extension of basic hygiene and sanitation all contributed to a steady lowering of the death rate, and although the primary purpose behind the West's newly intensified intervention in Southeast Asia was to advance its own economic interests, there was enough left over from the profits it made to pay for all the basic services mentioned above.

In the process of extending their rule over the whole of Southeast Asia, except for Siam which, at the eye of the storm, escaped with its essentials intact though stripped of its outer clothing, the Western powers concerned ended up by producing a new political map which, in reflecting the immovable realities of physical geography, bore at least a recognizable relationship to the one it replaced. But in effect the Southeast Asian states had been turned inside out, so that instead of looking inwards[16] they all faced out towards the central seas, which were effectively controlled by British naval power based upon the original three-port-axis. Thus the British, for the first time in history, had made Southeast Asia into a single functioning system, even though only a relatively small part of it was under their direct control.

From the foregoing manifestly Anglocentric account, however, one vital component is missing, namely the immigrant Asians, coming at first mainly from India but soon in far greater numbers from China. Once again, therefore,

---

[16]As noted above, this phrase calls for considerable qualification in respect of certain parts of Indonesia.

we come back to the old theme of Southeast Asia's relationship with its greater neighbours. Indeed the original purpose of the three-port-axis was not to carve out a new British sphere of influence in Southeast Asia (though admittedly that was what Raffles had wanted to do) but to promote the East India Company's trade with China. Nevertheless the forging of this new and much closer link between Southeast Asia and its two great neighbours had momentous consequences. From the foundation of Penang in 1786, Indian labourers, clerks, and traders played an indispensable part in its development, as later did many more of their countrymen in the other Southeast Asian territories under British rule elsewhere in Malaya and in Burma. But altogether more significant was the combined effect of British action in forcing China to permit free emigration and the simultaneous development of regular shipping services between the China coast, Hong Kong, Singapore, and Penang. Thus the minute trickle which was all that had hitherto escaped from the huge Chinese demographic reservoir suddenly swelled to a torrent, and an immense volume of pent-up human energy surged south to find new outlets in the underpopulated pioneer lands of the Nan Yang.

In most forms of employment, and more specifically within mining, agriculture, clerical work, commerce, and services of all kinds, the Chinese in the 19th century provided labour of temperate zone energy for Asian levels of remunieration. Without this, and in particular without the immense concentration of it in Singapore and the rest of the old British Malaya at the nerve centre of the whole system, the economic history of the past 150 years in Southeast Asia would present a far less spectacular story.

To say this, however, is by no means to say the last word on the subject. Long before the Second World War the presence of both alien Asian communities was increasingly resented by the indigenous Southeast Asian peoples, a fact which was deliberately exploited by the Japanese in their attempt to gain control over the immensely productive going concern of Southeast Asia, as indeed they did between 1942 and 1944. While the most obvious consequence of this interlude was that it indirectly precipitated the ending of Western colonial rule in Southeast Asia, the longer term effect may well prove to be that of bringing about a major economic realignment of the region towards Japan. For the mostly relatively small Southeast Asian countries still have little room for manoeuvre economically, and since for basically climate and geological reasons they tend to produce much the same commodities as one another, there is not a great deal of scope for closer mutual association based on inter-regional complementarity.

Meanwhile, however, many of these newly independent states have understandably tended to revert at least part way towards the introversion which was traditional in much of the region before the colonial era. In so doing they have become increasingly concerned over the extent to which their countries have come under the effective control of their non-indigenous minorities and this concern has already led to the exclusion of Singapore, 76 per cent of whose people are of Chinese origin, from the new Federation of Malaysia. But this very exclusion has in turn driven Singapore to more intensive manifestations of the dynamism which is the root cause of its neighbours' fears.[17] Thus the former problem of the hollow focus has taken on a new form, for in the middle of this central area there is now paradoxically a highly charged core.

Yet within the past few years hitherto unexpected developments have conspired to give a further twist to this situation. For not merely has a new problem arisen in the growing congestion in the shipping lanes through the Malacca Straits, but beneath the seas of the hollow centre the continental shelf is now thought to be an area of considerable and perhaps even great potential importance as a source of oil. So at the end of our survey we find that, despite all the relatively recent talk of air communications making Bangkok replace Singapore as the centre of Southeast Asia, the central sea area is in process of becoming more focal, and perhaps significantly less hollow, than ever before.

Whether all the various factors cited above make Southeast Asia what is sometimes called a major geographical region I do not propose to argue. What however seems clear to me is that, if not a unity, it is certainly an area with a great many problems in common, and all these problems are directly or indirectly rooted in its geography. It is on such grounds as these that I regard Southeast Asia as an intelligible field of area studies, and that I contend that geography is fundamental to area studies.

Geographical Literature relating to Southeast Asia

In trying to assess what geographers[18] have contributed to the study of the Southeast Asian area, one quickly becomes aware of a three-fold division between

---

[17] An essentially similar fear of Vietnamese (and particularly of North Vietnamese) dynamism, rather than ideology as such, is evident among their neighbours also.

[18] The term geographer is used here in its modern sense and does not therefore include old-style explorers or what the late Professor E. G. R. Taylor called "boy scout geographers."

the colonial period before 1942, the short but highly significant period of the Japanese occupation and its immediate and chaotic aftermath, and the post-colonial period since about 1950 (though the date varies from country to country owing to local circumstances) when field research again became possible.

Work Completed before 1942-43

Apart from some minor Japanese studies, all the geographical work published on Southeast Asia before 1942 was done by Westerners, and much the greater part of it by nationals either of one or other of the colonial powers with territories in the region, or of Germany which since the nineteenth century had played a leading role in geographical research in Asia. Among the colonial powers involved, the most impressive contribution was that of the French, who established a high quality of scholarship in their university at Hanoi (founded in 1917) which provided the base for much pioneer geographical research. However, the first study to mention, namely *L'Inde at l'Indochine* by Elisée Reclus, forming Tome 8 of his *Nouvelle Géographie Universelle*, 1883 (English translation 1884), considerably antedates this period. Although obviously the product of an earlier view of geography, Reclus's study is still an important work of reference, both on account of its penetrating insights and its wealth of detail about the crucial era of maximum geographical change in the late 19th century which is in danger of becoming the forgotten period between historical and contemporary geography. The next significant work to note is the contribution of two geographers, E. Chassigneux (already the author of *L'irrigation dans le delta du Tonkin*, 1912), and J. Brunhes, who respectively wrote sections on "Géographie" and "Géographie Humaine," in the first volume of *Un empire colonial française, L'Indochine*, edited by G. Maspéro (1923), and in 1929 J. Sion included an excellent overall study of "Indochine" in the second volume of his *Asie des Moussons* in the (new) *Géographie universelle* series under the direction of P. Vidal de la Blache and L. Gallois.

Nevertheless it was in the detailed pioneer research of two outstanding scholars, Charles Robequain and Pierre Gourou, that the French school made its greatest contribution to the study of Southeast Asian geography, beginning in 1929 with Robequain's doctoral thesis at Grenoble, *Le Than-hoa, étude géographique d'une province annamite*, which he followed with *L'Indochine française* (1939, English translation, 1944). Meanwhile Gourou, after first publishing *Le Tonkin* in 1931 showed a similar progression from detailed studies to wider but firmly based generalization, when he followed his masterly *Les pays du delta tonkinois* (1939, reissued 1966) with *L'utilisation du sol en Indochine française* (1940).

Besides these larger works which represent the core of the French contribution to the pre-war geographical study of Southeast Asia, there was an impressive total of over thirty articles (largely by professional geographers) on various aspects of Indochinese geography in French geographical serials between 1923 and 1942, and the French undoubtedly made the running, both qualitatively and quantitatively, in the Southeast Asian field at the last prewar International Geographical Congress at Amsterdam in 1938.

In comparison to the French, the British were late developers, though during the inter-war period some useful research was done by staff members in the new geography departments first at the University of Rangoon and later at Raffles College, Singapore. Indeed it is a strange but fortunate coincidence that two very different men, respectively the most prolific and, in my opinion, the most intellectually stimulating of modern British geographers, should each have worked for a few years in the University of Rangoon, and in so doing laid the foundations of geographical study in Burma. The first of these, the late L. Dudley Stamp, after having originally gone to Burma as an oil geologist in 1921, became Professor of Geology and Geography at Rangoon from 1923 to 1926, and on the basis of his work there wrote two articles, "Notes on the Vegetation of Burma"[19] and "Burma--An Undeveloped Monsoon Country"[20] as well as the best section (on Burma) in his compendious but very uneven *Asia*, first published in 1929 and still going strong. Following Stamp's work on physical geography, O. H. K. Spate, who went out to Rangoon in 1938, managed in the short time before the Japanese invasion to acquire an acute and deeply sympathetic perception of the underlying realities of the human geography of Burma. An indication of this can already be seen in his paper "The Beginnings of Industrialization in Burma"[21] and in another (with L. W. Trueblood) on "Rangoon: A Study in Urban Geography,"[23] but his best work on Burma appeared after the war brought his career in that country to an end in 1942.

Meanwhile in Singapore E. H. G. Dobby, another young British geographer

---

[19] *Geographical Journal*, Vol. 44 (1924), pp. 231-237.

[20] *Geographical Review*, Vol. 20 (1930), pp. 86-109.

[21] *Economic Geography*, Vol. 17 (1941), pp. 75-92.

[22] *Geographical Review*, Vol. 32 (1942), pp. 56-73.

who first arrived in Southeast Asia shortly before the outbreak of the Second World War, had begun the first significant geographical research in what was then called British Malaya. Out of this activity came his three papers "Settlement and Land Utilization in Malacca," "Singapore--Town and Country," and "Settlement Patterns in Malaya,"[23] before the extension of the war likewise interrupted his work in Malaya.

While few countries in the world offer such fascinating problems for geographical study as Indonesia, the pre-war Dutch contribution in this field, though certainly more extensive than that of the British in Burma and Malaya, was relatively meagre by comparison with the extremely impressive Dutch contribution in several cognate disciplines.[24] This situation reflected the fact that, although the Dutch had successively introduced university-level training in technology, law, and medicine in Java during the 1920's, it was not until 1940 that the process was extended to include the main arts and science subjects, so that no university teaching posts or research facilities in geography were available in Indonesia before the Second World War. Nevertheless S. Van Valkenburg served as a Geographer in the Netherlands Indies Topographical Service from 1921 to 1926, an experience which bore fruit in his articles on the economic geography of Java and on the agricultural regions of various parts of Southeast Asia after he had moved into the American academic orbit.[25]

Of the pre-war Dutch geographers based in the Netherlands and specializing, at least in part, on Indonesia, the most outstanding were H. Blink, G. Kuperus, C. Lekkerkerker, and L. Van Vuuren. In contrast to Blink, who wrote mainly on economic geography, and Kuperus who is best known for his detailed

---

[23]Published in Geographical Journal, Vol. 94 (1939), pp. 466-478; Geographical Review, Vol. 30 (1940), pp. 84-109; and Geographical Review, Vol. 32 (1942), pp. 211-232.

[24]Particularly relevant to geographers was the high quality cartography of the Netherlands Indies Topographic Service, a worthy monument of which remains the superb Atlas van Tropisch Nederland (Batavia, 1938). Mention should also be made of the excellent work of the geologists H. A. Brouwer, G. A. F. Molengraaff, L. M. R. Rutten, H. Staufer, J. H. F. Umbgrove, and R. W. van Bemmelen, the climatologists J. Boerema and C. Braak, the botanist G. G. J. van Steenis, and the soil scientist E. J. C. Mohr.

[25]See, for example, "Java: The Economic Geography of a Tropical Island," Geographical Review, Vol. 15 (1925), pp. 563-583; and "Agricultural Regions of Asia," Part 3 (Farther India), Economic Geography, Vol. 9 (1933), pp. 1-19, Part 8 (Malaysia,), ibid., Vol. 11 (1935), pp. 227-246, 325-337, Part 9 (Java) ibid., Vol. 12 (1936), pp. 27-44, and Part 10 (the Philippines), ibid., Vol. 12 (1936), pp. 231-249.

studies of small areas, notably Het cultuurlandschap van West-Soembawa (1937), Van Vuuren, whose period of publication extended from 1908 until into the 1940's, ranged comparably widely both geographically and topically, from Sulawesi to Sumatra, and from indigenous shipping to industrial raw materials and communism. But it was Lekkerkerker who in the 1930's really got to grips with what by then had become the central geographical problem of Indonesia, namely the steadily intensifying demographic congestion of Java. Thus in 1937 he produced a pioneer article, "Javas overbevolking,"[26] which was quickly followed by the first volume of his Land en volk van Java (1938). Indeed an almost obsessive concern with this same problem was apparent in the Dutch contribution to the section on colonial geography at the International Geographical Congress at Amsterdam in 1938, which included papers by G. Kuperus on "The relation between density of population and utilization of soil in Java," by M. B. Smits on "Population density and soil utilization in the Netherlands Indies" and by E. J. C. Mohr on "Relations between soil and population density in the Netherlands Indies."

Apart from their work on Java, however, the relatively few Dutch geographers specializing on Indonesia were so thinly spread over the vastness of its territories that they could do little more than touch the fringe of its multitudinous geographical problems. Thus while Van Vuuren contributed a useful summary of "Geographical aspects of the Netherlands Indies" to L. M. R. Rutten's Science and Scientists in the Netherlands Indies (1932), it would probably have been premature at that stage to attempt a more substantial geographical appreciation of the country as a whole.

However, besides the Dutch scholars referred to above, several of their German contemporaries made very substantial contributions to the study of Indonesian geography, most of them with reference to areas outside Java. Thus K. Helbig made several important cultural geographical studies on Sumatra and such nearby islands as Bangka and Nias, and also wrote two books on Kalimantan, while M. Kornrumpf specialized on Sulawesi. Meanwhile H. Lehmann, in addition to writing on Sumatran population, also contributed articles on Indonesian cities and related topics.

Nevertheless it was in respect to Thailand and the Philippines that the prewar German contribution to Southeast Asian geographical studies was most

---

[26] Tijdschrift van het Konninklijk Aardrijkskundig Genootschap, Vol. 54 (1937), pp. 866-886.

decisive for, notwithstanding a university tradition going back to 1917 in the former of these countries and even to 1595 in the latter,[27] neither had developed any local centres of geographical research before 1941. Thus apart from two outstanding German studies, namely Wilhelm Credner's Siam, das Land der Thai (1935) and Albert Kolb's Die Philippinen (1942), each of which had been preceded by several articles based on these respective authors' local field work, I have been unable to trace any published work by a professional geographer on either Thailand or the Philippines prior to Stamp's paper "Siam before the War,"[28] though some substantive work of great value to geographers was meanwhile completed by American experts in related disciplines, including J. M. Andrews, C. C. Zimmerman, and R. L. Pendleton.

Although Credner's and Kolb's two masterpieces have inevitably dated in certain respects, they indubitably remain among the small handful of really satisfying geographical studies of individual Southeast Asian countries. Had it not been for them, our knowledge of Southeast Asia would have remained disastrously inadequate until well after the Second World War. All in all, therefore, the German contribution to the pre-war geographical study of Southeast Asia surpassed that of any of the colonial powers with dependencies in the region, and it is not therefore surprising to find that it was another (at that time) German geographer, Karl J. Pelzer, who first used the name Southeast Asia in the title of a geographical monograph, namely his aforementioned Die Arbeiterwanderungen in Sudestasien (1935). For, as this account of pre-war geographical work on Southeast Asia shows only too clearly, colonial blinkers had prevented all but the non-colonial Germans from ever attaining a proper awareness of the totality of Southeast Asia as more than the sum of a series of separate colonial dependencies. What of course brought such an awareness, not merely to Western geographers but also to the educated Western public at large, was the Japanese seizure of the whole of what they had come to call "the southern resources area" and its inclusion within their Greater East Asian Co-prosperity Sphere, a development which as one of its minor consequences inevitably put a stop for several years to normal geographical research within that area.

---

[27] The University of San Carlos in Cebu City was founded in 1595 and that of Santo Tomas in Manila followed in 1611. Some 25 other universities have been added since 1900.

[28] Geographical Journal, Vol. 99 (1942), pp. 209-224.

Work Completed between 1942-43 and 1945-50

Although this great interruption represented a tremendous setback to basic geographical research and even more to the training of indigenous Southeast Asian geographers which had just begun to make significant progress in various parts of the region, it had at least one beneficial effect. For the insatiable demand for more information about this now critical part of the world provided many geographers with the stimulus and/or the opportunity which led them either to complete for publication work on Southeast Asia on which they had already done substantial research or, in the case of some who had not previously worked in this area, to apply their basic geographical skills and other relevant expertise (such as appropriate linguistic competence) to working over the immense amount of raw material in geological, agricultural, census, trade, and other reports in order to produce so-called secondary studies.[29]

While obviously field work is fundamental to geography, a prolonged absence from it from time to time is no bad thing. Like the imminent prospect of death, deprivation of access to one's chosen part of the world may prove--at least to the geographer--a wonderful means of concentrating the mind. No better proof of this can be found than in the Western geographical literature on Southeast Asia which appeared during and immediately following the period of the Japanese occupation. As examples I would begin by citing Spate's attractive and lively little paperback Burma Setting (1943) and his perceptive article "The Burmese Village,"[30] and also Leonard Unger's basic geographical study on "The Chinese in Southeast Asia."[31] A larger and more substantial work was The Economic Development of the Netherlands Indies (1942) by J. O. M. Broek who had already written an excellent article on "The economic development of the outer provinces of the Netherlands Indies"[32] and proceeded to follow it with several more, of which "Diversity and unity in Southeast Asia"[33] was a brilliant pioneer contribu-

---

[29] In fact of course many works categorized by this unfortunate term are of primary importance.

[30] Geographical Review, Vol. 35 (1945), pp. 523-543.

[31] Geographical Review, Vol. 34 (1944), pp. 196-217.

[32] Geographical Review, Vol. 30 (1940), pp. 187-200.

[33] Geographical Review, Vol. 34 (1944), pp. 175-195. Although by this time Broek was working in the United States, his interest in Southeast Asia owed much to his earlier training in the Netherlands, whither he returned for a time after the Second World War as Professor of Social Geography in Utrecht.

tion to the geographical concept of Southeast Asia. This brief and selective list is fittingly crowned with Karl Pelzer's masterpiece Pioneer Settlement in the Asiatic Tropics (1945), though two very interesting articles which appeared subsequently must also be mentioned, namely E. H. G. Dobby's "Some aspects of the human ecology of Southeast Asia"[34] (1946) and Wilbur Zelinsky's "The Indochinese peninsula: a demographic anomaly" (1950).[35]

Short though this list is, the works cited introduced many new and thought provoking ideas, such as the analogy between Southeast Asia and the Balkans, the definition of the region as the lands bordering the South China Sea, and the characterization of it as a low density zone between India and China in which the Chinese had played a variety of roles comparable to those of European pioneers in North America. All of these, and many other ideas which have played a seminal part in the growth of the modern geographical concept of Southeast Asia, date from this brief but formative interlude, which happens to coincide with the period of my own first encounter with Southeast Asia. By the simple device of volunteering for service in the Middle East in March 1941 I arrived in Singapore in the following August, a few months ahead of the Imperial Japanese Army. Compounding my recklessness by volunteering for a P.O.W. working party destined for Japan in mid 1942, I was soon consigned to railway construction in Thailand and so began a three year spell of somewhat unorthodox field study over a wide area of that country. Apparently I was destined to specialize in Southeast Asia, for though I had never heard the name mentioned before I went into captivity, I came out to find that Southeast Asia was in process of arriving. The experience was rather like that of the late Dean Acheson in being present at the creation.

Work by Western Scholars Completed since 1950

The period since 1950, by which time work based on post-war field research was beginning to be published, is the most difficult of all to assess, pri-

---

[34] Geographical Journal, Vol. 108(1946), pp. 40-54.

[35] Far Eastern Quarterly, Vol. 9 (1950), pp. 115-145. It is appropriate here to note that two valuable Geographical Handbooks on Southeast Asian countries were produced by the (British) Naval Intelligence Division during the war, namely No. 510, Indo-China (1943) and Nos. 518 and 518A, Netherlands East Indies (1944). These composite studies, which appeared under the general editorship of H. C. Darby, owed much to two other British geographers, J. C. Stuttard and S. H. Beaver.

marily because of the problem of keeping abreast of the much larger volume of work now emanating from an ever growing number of centres and individuals.

By 1950 many Western scholars had either resumed or begun active research in Southeast Asian geography and many of the more serious gaps began at least in part to be filled. Perhaps the most obvious of these hitherto had been that of a basic university textbook covering the whole area and this was now provided by Dobby's Southeast Asia (1st edition 1950). Inevitably this drew heavily on earlier geographical studies, most of which have been listed above, as well as on relevant publications in related fields. Undoubtedly also Dobby was at his best on Malaya, on which he had already written several papers, and continued to produce more, largely filling out the regional details of the peninsula. But the supreme contribution of his book was that, by providing the first overall geographical study of Southeast Asia as an entity, it made it possible for university teachers in many parts of the world to introduce major geographical courses on this region. Thus in spite of what Spate described as a style which constantly sets one's teeth on edge, Dobby's book has probably done more to establish the geographical study of Southeast Asia than any other single publication. About the same time, having already completed a paper on "The Thailand-Burma railway,"[36] I began to make my own first attempt at an overview of the region from the standpoint of political geography, which was published as part of The Changing Map of Asia, edited by W. G. East and O. H. K. Spate, the first edition of which also appeared in 1950.

Shortly afterwards some of the bigger gaps in regional geography received attention, most notably by the work of a series of American geographers, led by Joseph E. Spencer, on the Philippines. Thus Spencer published his Land and People in the Philippines in 1952 and subsequently collaborated with F. L. Wernstedt in a monumental work, The Philippine Island World (1967), while other useful contributions on the same area were R. E. Huke's collection of studies Shadows on the Land (1963) and Alden Cutshall's The Philippines: Nation of Islands (1964). In addition, all of these American scholars published articles on various aspects of Philippine geography and others to note in this connexion include McIntyre, Ullman, and Vandermeer.[37]

---

[36] Economic Geography, Vol. 23 (1947), pp. 85-89.

[37] Mike McIntyre, "Geographical Regionalism in Leyte," Philippine Geographical Journal, Vol. 3 (1955), pp. 31-39 and "Geographical Regionalism in Samar," ibid., Vol. 4 (1956), pp. 7-13; Edward L. Ullman, "Trade Centers and Tributary Areas of the Philippines," Geographical Review, Vol. 50 (1960), pp. 41-55; C. Vandermeer, "Population Patterns on the Island of Cebu," Annals of the Association of American Geographers, Vol. 57 (1967), pp. 315-337.

Meanwhile in post-war Indonesia two Dutch geographers, F. J. Ormeling and H. Th. Verstappen, both of whom served for a time in the Topographical Service of the Indonesian Army, made significant contributions to research in social geography, notably The Timor Problem (1955), and in the geomorphology of Sumatra and other parts of the archipelago. Other post-war geographical work on Indonesia by Dutch scholars includes articles by Pannekoek on the geomorphology of Java, Breman, Hanrath, and Tennekes on various aspects of population geography, Winsemius on West Irian, and a most useful recent monograph by Lagendijk of Utrecht on Gegesik Wetan, a rice growing village in the coastal plain of northern Java.[38]

Likewise in the former French Indochina there has during the past fifteen years been a notable revival of geographical research by French scholars, though, apart from a brief historical study of the origins of Hanoi by Azembre,[39] it has for understandable reasons concentrated on areas other than North Vietnam. This post-war French research in Indo-China, and also, of late, in other parts of Southeast Asia, has been largely stimulated by the Centre d'Etudes de Géographie Tropicale (CEGET) directed by Professor Guy Lasserre of the University of Bordeaux, and much of it has been published in that institution's journal Cahiers d'Outre-Mer. Nevertheless the most important individual contribution is that of Jean Delvert, who worked for some years in Phnom-Penh before becoming Professor of Regional Geography at the Institut de Géographie at the Sorbonne, and whose magnum opus, Le Paysan Cambodgien (1961), continues the tradition established by Gourou's 1936 study of the peasants of the Tonkin delta.

Other recent French work on Cambodia includes Delvert's study on fishing, articles by two Staff members of the new University of Phnom Penh, R. Blanadet and J. Migozzi, respectively on agricultural-regional and population

---

[38] A. J. Pannekoek, "Geomorfologische waarnemingen op het Djampang-Plateau in West-Java,"Tijdschrift van het Koninklijk Aardrijkskundig Genootschap, Vol. 63 (1946), pp. 340-367; J. C. Breman, "Java: Bevolkingsgroei en demografische structuur," ibid., Vol. 80 (1963), pp. 252-308; J. J. Hanrath, "Transmigratie in Indonesie in het algemeen en in Celebes in het bijzonder," Tijdschrift voor Economische en Sociale Geografie, Vol. 45 (1954), pp. 133-138; J. Tennekes, "De berolkingspreiding der residentie Besoiki in 1930," Tijdschrift van het Koninklijk Aardrijkskundig Genootschap, Vol. 80 (1963), pp. 309-423; J. Winsemius, "Een algemeen planologisch schema voor Nieuw-Guinea's ontwikkeling," ibid., Vol. 72 (1955), pp. 118-129; J. M. G. Lagendijk, Gegesik Wetan, een rijstdorp in de noordjavaanse Kustvlakte (Utrecht: Geografisch Instituut, 1971).

[39] G. Azambre, "Origines de Hanoi," Bulletin de la Société des Etudes Indochinoises, Vol. 33 (1958), pp. 263-300.

geography, and an invaluable monograph on industrialization by Kamelgarn.[40] Here also it is appropriate to mention Robert Garry, a former member of the French Colonial Service in Indochina, and more recently a Professor of Geography at the Université deMontréal, whose writings on Cambodian subjects range from pepper cultivation to geopolitics.[41] Meanwhile new geographical ground has been broken in Laos by Taillard of the Centre National de la Recherche Scientifique (CNRS),[42] Lafont has written on swidden cultivation in the uplands of central Vietnam,[43] and between 1961 and 1966 R. Teulières published several papers on various aspects of the economic geography of South Vietnam.[44]

Similarly in Malaya and Singapore British scholars dominated geographical research during the first decade after the Second World War. Under E. H. G. Dobby, who returned to Singapore and became the first Professor of Geography in the new University of Malaya (created in 1949, through the merger of the former Raffles College and the King Edward VII School of Medicine), rapid progress was made in building on the foundations he had already laid before the War. An outstanding achievement in this connexion was the launching in 1953 of the Malayan Journal of Tropical Geography which rapidly and deservedly won international recognition, a fact which owed much to the numerous contributions both of Dobby himself and of the new Staff members whom he and his successors recruited.[45] Virtually all the post-war British geographers who have served in the University of Malaya and/or its successors have made significant contribu-

---

[40]Rachel Kamelgarn, "Les industries au Cambodge," Etudes Asiatiques, 1967, pp. 43-165.

[41]"The Changing Fortunes and Future of Pepper Growing in Cambodia," Journal of Tropical Geography, Vol. 17 (1963), pp. 33-42, and "La géopolitique du Cambodge et la deuxième guerre d'Indochine," Etudes de géographie tropicale offertes à Pierre Gourou (Paris and The Hague: Mouton, 1972), pp. 165-182.

[42]Ch. Taillard, "L'Irrigation dans le Nord du Laos: L'exemple du bassin de la Nam Song à Vang Vien," in Etudes de géographie tropicale offertes à Pierre Gourou, pp. 241-256.

[43]P.-B. Lafont, "L'Agriculture sur brûlis chez les Proto-Indochinois des hauts plateaux du Centre Vietnam," Cahiers d'Outre-Mer, 1967, pp. 37-48.

[44]See, for example, "Le thé au Sud-Vietnam," Cahiers d'Outre Mer, 1961, pp. 182-209; "La culture du caféier au Sud-Vietnam," ibid., pp. 293-316; "Le tabac au Vietnam du Sud," ibid., 1966, pp. 174-186; and "L'Evolution récente de la riziculture au Vietnam du Sud," ibid., pp. 382-398.

[45]The name of this Journal was changed to the Journal of Tropical Geography in 1958.

tions to local research. Among these are five books, respectively by B. W. Hodder, Man in Malaya (1959), by Paul Wheatley, The Golden Khersonese (1961) and Impressions of the Malay Peninsula in Ancient Times (1964), and two by J. C. Jackson, Planters and Speculators--Chinese and European Agricultural Enterprise in Malaya 1786-1921 and Sarawak--A Geographical Survey of a Developing State, both in 1968.[46] In addition, a most impressive range of papers appeared by Dobby, mainly on agricultural and regional geography, by I. Douglas on geomorphology and by W. L. Dale on climatology, by D. W. Fryer, G. R. Elliston, J. C. Jackson, and R. F. Kinloch on economic geography, by B. W. Hodder and W. D. McTaggart on social and political geography, and by P. Wheatley and J. C. Jackson respectively on early and more recent historical geography. Of the British geographers who, though not based in one or other of the local universities also worked on Malaya-Singapore during the same period, Jean Carter has published on coastal geomorphology, P. P. Courtenay on economic geography, and myself on political geography.

However by the late 1950's the domination of Southeast Asian geography by nationals of the former metropolitan powers was breaking down, a process which has operated in at least three inter-related ways. First, thanks to the Commonwealth connexion and the increasing international use of the English language, the focally situated and well-provided-for universities of Malaya (then including Singapore) came to attract well qualified expatriate staff members from such countries as Austria, Ceylon, Hong Kong, the Netherlands, and New Zealand; secondly, many American, British, French, and also Indian and Japanese geographers began to emulate their German colleagues in ranging more widely over Southeast Asia; and thirdly, and potentially most significant of all, Southeast Asian geographers themselves are now playing a rapidly increasing role in both teaching and research within the region. While the first two of these developments will be treated relatively briefly here, the third is of such importance as to call for a section on its own.

Of the non-metropolitan expatriates working in the universities of Malaya and Singapore, several have published significant papers on local

---

[46] Professor Hodder is now at the School of Oriental and African Studies, University of London, Dr. Jackson at the University of Hull, and Professor Wheatley at the University of Chicago.

subjects, respectively in geomorphology,[47] in climatology,[48] agricultural geography,[49] and social-urban-political geography.[50]

In the second category the pioneer example is provided by Charles Robequain who shortly after the Second World War extended the range of his studies from French Indochina to maritime Southeast Asia and so came to write Le Monde Malais (1946), which was later translated into English as Malaya, Indonesia, Borneo and the Philippines (1954). Although this contained many penetrating flashes of insight, it must be admitted that, based as it was on a relatively superficial acquaintance of the areas concerned, it fell distinctly below the level of his earlier studies, and it is particularly sad that this should have been the last major work before his premature death in the early 1960's of one of the greatest geographers who has ever worked in the Southeast Asian field. More recently several other French geographers have begun to work in areas of Southeast Asia outside Indochina, including once again Delvert who has published both on Thailand and on Indonesia,[51] and Bruneau and Lainé also on various parts of Thailand.[52]

---

[47] See, for example, R. J. Eyles, "A Classification of West Malaysian Drainage Basins," Annals of the Association of American Geographers, Vol. 61 (1971), pp. 460-467; R. D. Hill, "Changes in Beach Form at Sri Pantai, Northeast Johore, Malaysia," Journal of Tropical Geography, Vol. 18 (1964), pp. 19-27; J. J. Nossin, "The Geomorphic History of the Northern Pahang Delta," ibid., Vol. 20 (1956), pp. 54-64; and S. B. St. C. Swan, "Maps of Two Indices of Terrain, Johor, Malaysia," ibid., Vol. 25 (1967), pp. 48-57.

[48] E.g., S. W. Nieuwolt, "The Urban Microclimate of Singapore," Journal of Tropical Geography, Vol. 22 (1966), pp. 30-37, and "Diurnal Rainfall Variation in Malaya," Annals of the Association of American Geographers, Vol. 58 (1968), pp. 313-326.

[49] R. D. Hill, "Rice Cultivation Systems in Malaya," World Crops, Vol. 18 (1964), pp. 10-14.

[50] E.g., T. G. McGee, "Aspects of the Political Geography of Southeast Asia: A Study of a Period of Nation Building," Pacific Viewpoint, Vol. 1 (1960), pp. 39-58; idem, The Southeast Asian City (London, 1967); and T. G. McGee and W. D. McTaggart, Petaling Jaya: A Socio-Economic Survey of a New Town in Selangor, Malaysia, Pacific Viewpoint Monograph No. 2 (1967).

[51] J. Delvert, "Note sur la 'Plaine de Bangkok' et la zone pionnière de Kanchanaburi (Thailande)," Bulletin de l'Association de Géographes Français, 1968, pp. 2-11 and L'Indonésie (Paris, 1964).

[52] E.g., M. Bruneau, "Un Village du Nord de la Thailande: Ban Muang Wa," Cahiers d'Outre Mer, Vol. 23 (1970), pp. 140-174, and J.-P. Lainé, "Bangkok: problèmes de croissance urbaine," ibid., Vol. 24 (1971), pp. 394-409.

Likewise among the British, D. W. Fryer, J. C. Jackson, and myself have each published fairly extensively on Indonesian topics,[53] and Ronald Ng, a British citizen of Hong Kong origin, has written several papers on both the population and agricultural geography of Thailand.[54] Of the post-war German geographers active in Southeast Asian research the most productive is Professor Harald Uhlig of Giessen, with contributions on both West and East Malaysia and Cambodia and others on more widely ranging regional topics.[55] Meanwhile Professor Kolb of Hamburg, though shifting his primary focus of interest farther to the north, has nevertheless included a chapter on Vietnam in his magnificent Ostasien (1967), English translation East Asia (1971), and also found time to write a stimulating short study Sudostasien im Weltbild unserer Zeit for the 30th German Geographentag in 1957. In addition Klaus Hausherr has produced pioneer social and cultural geographical studies of areas in Mindanao, K. Horstman has written on population problems, industrialization, and economic development in Indonesia, Küchler on the cultural landscape of Penang, Reiner on the Moluccas, and Voss on aspects of the physical geography of the Philippines.[56] Finally among German-language publications reference should be made to two substantial studies on the economic geography of Thailand by the Austrian scholar Ernst Troger,[57] and to another on Indonesia

---

[53] See, for example, Fryer's Indonesia (Croydon, 1957, No. 5 of Longmans Australian Geographies); Jackson's Sarawak: A Geographical Survey of a Developing State (London, 1968); and my "Indonesia: A Giant Astir," Geographical Journal, Vol. 138 (1972), pp. 154-165.

[54] E. g., "Population Explosion in the North Thai Hills," Geographical Magazine, Vol. 43 (1971), pp. 255-263.

[55] See, for example, his "Die Volksgruppen und ihre Gesellschafts- und Wirtschafts-entwicklung als Gestalter der Kulturlandschaft in Malaya," Mitteilungen der Österreichischen Geographischen Gesellschaft, Vol. 105 (1963), pp. 5-94, and "Südostasien--Länder, Völker, Wirtschaft," Information zur Politischen Bildung, Vol. 148 (1972), pp. 1-32.

[56] J. Küchler, Penang--Kulturlandschaftswandel und ethnisch-soziale Stuktur einer Insel Malaysias, Giessener Geographische Schriften, Vol. 13 (1968); E. Reiner, Die Molukken, Petermanns Geographische Mitteilungen, Ergänzungsheft 260 (1956); F. Voss, "Typische Bergflächenformen tropischen Kiegelkarstes auf den Philippinen," Geographische Zeitschrift, Vol. 55 (1970), pp. 214-227.

[57] "Ban-Pae-Lungar: Studie zur Siedlungs- Wirtschafts- und Bevölkerungsgeographie eines Dorfes in Nordthailand," Mitteilungen der Österreichischen Geographischen Gesellschaft, Vol. 102(1960), pp. 165-198, and "Struktur und Probleme der Wirtschaft Thailands," ibid., Vol. 104 (1962), pp. 270-336.

by his fellow countryman H. Küpper.[58]

Among American geographers working in Southeast Asia outside the Philippines, the most notable post-war contribution has come from Karl Pelzer whose thoroughly researched papers cover a wide range of subjects, though his most substantial and eagerly awaited piece of research, on northeastern Sumatra, has not yet been published. Besides Pelzer, W. A. Withington has also written extensively on Sumatra,[59] J. O. M. Broek continued his Indonesian studies,[60] and more recently J. L. Cobban has written a most interesting paper on the historical geography of Jakarta.[61] Meanwhile Norton Ginsburg has worked on regional aspects of urbanization and planning[62] and was the senior collaborator (with C. F. Roberts) in a study of Malaya (1958). Other American work includes that of Blaut on Singapore agriculture, Sternstein on various aspects of Thailand, and Huke similarly on Burma.[63] And special mention should be made of the most attractive study Thailand, Aspects of Landscape and Life by R. L. Pendleton assisted by Robert C. Kingsbury and others (1962).

Finally, within the English speaking world it remains to mention the papers of the Canadian J. E. Stenton, respectively on the Mekong project, Malaysia, and the Philippines, and the rather more numerous contributions of Australian and New Zealand geographers. Among the former Chapman has undertaken important field research in Thailand, only a part of which has yet been published,[64]

58"Indonesien: Entwicklungstendenzen im Malayischen Archipel," ibid., Vol. 105 (1963), pp. 95-118.

59E.g., "Problems and Potentialities of Development of Western Indonesia: Sumatra as an Example of Third World Characteristics," Journal of South-East Asia and the Far East, 1969, pp. 79-103.

60"East Indonesia: Problems and Prospects," Far Eastern Survey, Vol. 20 (1951), pp. 65-67; "Resources of the Tropics III: Southeast Asia," Focus, Vol. 4, No. 2 (1954); and "Indonesia," ibid., Vol. 7, No. 4 (1956).

61"Geographical Notes on the First Two Centuries of Djakarta," Journal of South-East Asia and the Far East, Vol. 44 (1971), pp. 108-150.

62"The Great City in Southeast Asia," American Journal of Sociology, Vol. 40 (1955), pp. 455-462.

63J. M. Blaut, "The Economic Geography of a One-Acre Farm on Singapore Island," Malayan Journal of Tropical Geography, Vol. 1 (1953), pp. 37-48; L. Sternstein, "Contemplating a Hierarchy of Centres in Thailand," Pacific Viewpoint, Vol. 7 (1966), pp. 229-235; and R. E. Huke, "Burma," in Alice Taylor (ed.), Focus on Southeast Asia (New York, 1972), pp. 63-81.

64E. C. Chapman, Thailand (Croydon, 1965), No. 23 of Longmans Austalian Geographies.

Rutherford has written on double cropping of rice in Penang,[65] and Missen has recently produced an important book Viewpoint on Indonesia: A Geographical Study (1972), which is particularly concerned with developmental problems. Mention should also be made here of O. H. K. Spate's attractive paper on Mandalay and Rangoon, which most appropriately formed his contribution to the memorial volume to Sir Dudley Stamp published in 1968. Besides R. D. Hill and R. J. W. Neville, whose work has already been mentioned, New Zealand geographers who have worked on Southeast Asia include Marion Ward, mainly on Malayan ports,[66] and Iain Buchanan, the author of Singapore in Southeast Asia (1972).

To complete this section on post-war academic work by Western geographers it remains to note the Southeast Asian sections in four works of extensive coverage, and a further four books dealing specifically with Southeast Asia as a whole from the various standpoints indicated by their respective titles. Within the first group, which comprises Karl Helbig, Am Rande des Pazifik (1949), Pierre Gourou, L'Asie (1953), J. E. Spencer, Asia East by South (1954, second edition J. E. Spencer and W. L. Thomas, 1971), and Norton S. Ginsburg (ed.), The Pattern of Asia (1958), the last two probably have the most useful contributions to offer to the Southeast Asia specialist, though this is largely a matter of personal preference. In the second group, T. G. McGee's The Southeast Asian City (1967), provides a brief but welcome survey of a critical subject, and Keith Buchanan's The Southeast Asian World (1967), is a highly readable, stimulating, and controversial short introduction to a very long subject. Possessing similar merits but also providing considerably more data in support of a different view of the economic problems of the region is D. W. Fryer's Emerging Southeast Asia (1970), which is clearly based upon wide experience and long sustained research. Finally, as regards my own South-East Asia--A Social, Economic, and Political Geography (1964), I will content myself by remarking that this overweight volume is a faithful reflection of its author.

Since the foregoing account is concerned specifically with academic--or what is sometimes curiously called pure--geography, it is necessary to add that

---

[65] J. Rutherford, "Double Cropping of Wat Padi in Penang, Malaya," Geographical Review, Vol. 56 (1966), pp. 239-255.

[66] E.g., "Port Swettenham and Its Hinterland, 1900-1960," Journal of Tropical Geography, Vol. 19 (1964), pp. 19-78.

in recent years several Western geographers have been deeply involved in the application of geography to the solution of practical problems in Southeast Asia, and although only a small proportion of their work is published except in official and/or private reports, its importance may in the longer term be of lasting academic as well as practical importance. Outstanding in this field has been the work of Gilbert F. White for the ECAFE Mekong Committee, as referred to in his paper "Contributions of geographical analysis to river basin development" and in the report by White, De Vries, Dunkerley, and Krutilla to that Committee in 1962. In this connexion moreover, the magnificent Atlas of Physical, Economic and Social Resources of the Lower Mekong Basin (United Nations, 1968), provides an excellent example of the kind of academic geographical spin-off which may result from such activity. Perhaps it is also pertinent to note that at least one professional geographer, R. F. Kinloch, was for several years on the Staff of the ECAFE Mekong Annex in Bangkok, and that the Department of Geography of the School of Oriental and African Studies in London is now carrying out an investigation under ECAFE Mekong auspices[67] of changing land use and agricultural innovation in the Lam Pao Tributary Project Area of Northeast Thailand.

Within Southeast Asia itself the Department of Geography in the University of Singapore is making a particularly wide ranging contribution to applied research on matters of local concern, including air pollution, the effects of coastal reclamation, physical oceanography, and the problem of oil spills and oil pollution, and also various aspects of planning, low-cost housing, residential mobility, and the hawker problem.

These particular cases have been cited merely as examples of the kind of applied geographical work which is being carried out in many parts of Southeast Asia. But in this, as also in purely academic geography, the active professional participants no longer consist solely of Westerners for more and more well qualified Asian geographers are now becoming available.

The Contribution of Asian Geographers to Southeast Asian Studies

In more senses than one Japanese geographers specializing on Southeast Asia occupy an intermediate position between their Western and their Southeast

---

[67] With financial support from the British Overseas Development Administration and the Social Science Research Council. More recently a team from the University of Michigan, under the direction of Professor L. A. P. Gosling, has begun a study of population resettlement in the Pa Mong project area, also under the auspices of the ECAFE Mekong Committee.

Asian counterparts. Such Japanese geographical research on Southeast Asia as was undertaken before the Second World War seems to have been at least indirectly related to Japan's imperial aspirations in that region, and some professional geographers gained extensive experience there during the occupation period. However, very little of their work has been published, and hardly any of it, so far as I know, in any Western language. During the first immediate post-war decade Japanese scholars had virtually no opportunity and apparently also little inclination to specialize on Southeast Asia, but thereafter a change became evident and at the I. G. U. Regional Conference in Japan in 1957 a Symposium to which five Japanese (K. Aki, H. Arakawa, A. Bekki, K. Iizuka, and F. Ueno) as well as one Indian (S. B. Chatterjee), one Frenchman (P. Gourou), one American (A. Cutshall), and one Englishman (E. H. G. Dobby) contributed short papers which were included in the Proceedings, published in 1959.

At the I. G. U. Regional Conference in Malaya in 1962 Japanese geographers were present in great strength but only one of them contributed a paper (unpublished) on a Southeast Asian topic. Since then more Japanese geographical research has been undertaken on various parts of Southeast Asia, among others by A. Bekki on Indonesia, K. Kikuchi on Indochina, Ota on Malaysia, and M. Murakami on Singapore. Virtually all of their published work is in Japanese though in some cases, as with Murakami's interesting study of the industrialization of Singapore, 1968, an English abstract is provided. However, under the auspices of two extremely impressive multi-disciplinary research institutions, namely the Institute of Developing Economies, Tokyo, and the Center for Southeast Asian Studies, University of Kyoto, both of which provide funds for extended field work and subsequent publication in English, some very important Japanese geographical work on Southeast Asia has recently become internationally accessible. An outstanding example of the work sponsored by the Institute of Developing Economies is a masterly monograph Land and Peasants in Central Luzon (1969), by the geographer A. Takahashi who has since become an Associate Professor of International Economics at Tokyo University, and an interesting shorter contribution by K. Iizuki, "An approach to Asian Studies from the standpoint of Human Geography," was published in the Institute's journal The Developing Economies in 1944. At the Kyoto Center of Southeast Asian Studies team work is much to the fore though the research of team members is published under the names of the individuals concerned. In this connexion most of the strictly geographical work is in the application of physical geography to agricultural needs, as in the three papers by Y. Takaya on the rice lands of the Chao

Phraya lowlands (1969-71) and a related emphasis is also evident in T. Tomosugi's invaluable contribution on the historical development of irrigation and drainage in the Chao Phraya delta, 1966. In addition to the work done by professional geographers at the Kyoto Center, much of that by their colleagues in such related fields as agricultural economics is also of great value to geographers. Besides Thailand, various parts of Indonesia and other Southeast Asian countries have already been studied and early in 1973 a strong team, again including Takaya, has begun work in Burma. Finally among Japanese geographical contributions on Southeast Asia mention should be made of F. Ueno's study of rural settlements and rice harvests in Java, which has been included in R. L. Singh's collection Rural Settlements in Monsoon Asia, Essays in Honour of Professor Jiro Yonekura (1972).

By comparison with the Japanese, the other Asian neighbours of Southeast Asia have shown relatively little geographical research interest in the region, apart from Professor Nafis Ahmad, formerly of Dacca, who has published a useful article "Planning the economic development of Burma,"[68] and two Indian scholars, J. N. Bhatta and M. Halim Khan who have made valuable contributions to the establishment of geography--both pure and applied--in post-war Indonesia.

In considering the way in which Southeast Asian scholars have begun to take the initiative in geographical teaching and research within their own countries it is perhaps best to begin by looking again at Indochina which in an ironic way appears to illustrate the saying that the first shall be last. For although as already noted, it was in Vietnam, and more specifically in Hanoi, that the most advanced centre of geographical research had been established in colonial times, the conditions created by the prolonged Vietnam War have proved disastrous to a subject which more than any other lives by its field work, and although important progress has meanwhile been made in Cambodia, this has so far been almost entirely the work of a new generation of French geographers.

In Burma the pre-war staff of the University of Rangoon had already trained a small group of geographers who, under the determined and inspiring leadership of Professor Daw Thin Kyi, have maintained a high standard of teaching throughout the postwar period. But, besides its dependence on field work, geographical research cannot indefinitely flourish in a state of intellectual isolation from the outside world, such as has existed in Burma for many years. In all the circumstances, therefore, the three publications, respectively by Maung

---

[68] Oriental Geographer, Vol. 14 (1970), pp. 1-33.

Kyaw (1954), Saw Hanson Tadaw (1959), and by Daw Thin Kyi and Daw Kyin Htay (1968), though certainly small in number, should be seen as an indication of the determined attempt of these Burmese geographers to keep their subject alive and in a fit state to make a fuller contribution when circumstances permit it to do so.

By comparison with Indochina and Burma, Malaya and Singapore suffered no serious interruption to the continuity of normal academic life except during the period 1942-1945. Thus, with a growing number of well-trained local geography teachers in the schools, the academic quality of many of their students entering the University of Malaya by the 1950's was encouragingly high, and several of the ablest graduates, usually after further training in Britain, were recruited to the staffs of the University of Malaya's geography departments in its two divisions at Kuala Lumpur and Singapore. By the time these two institutions became separate universities in their own right in 1962 a majority of the members of their geography departments, respectively under Professor Robert Ho and Professor Rudolph Wikkramatileke (of Ceylon), were Asians, all but two of them from Malaya-Singapore.

As a result of particular happy timing, the holding of the 1962 I. G. U. Regional Conference at Kuala Lumpur provided an opportunity to demonstrate to their colleagues from many parts of the world what this new generation of Southeast Asian geographers could do. As one who attended a full share of the exceptionally well organized programme of lectures, discussions, and excursions at this most enjoyable conference, I still recall with amusement the surprised and in some cases totally mystified reaction of many of my fellow Westerners to the fact that such things could be, though fortunately the splendid _Guide to Conference Tours_ and the subsequent publication of two special issues of the _Journal of Tropical Geography_ devoted entirely to a selection of conference papers provided tangible and permanent confirmatory evidence.

Since then the geography departments at Kuala Lumpur and Singapore have grown considerably. Professor Ooi Jin Bee has taken over at Singapore from Professor Wikkramatileke (now at the University of Victoria, British Columbia), and although no permanent successor to the late and greatly missed Professor Ho has yet been appointed at Kuala Lumpur, Tunku Shamsul Bahrin has been serving as Acting Head of a staff which, except for occasional visiting members, is now entirely Asian. Meanwhile Dr. Hamzah Sendut, previously on the geography staff at Kuala Lumpur, has become Vice Chancellor of the new University at Penang (Universiti Sains Malaysia) where geography is being introduced, and

Dr. K. S. Sandhu, formerly of the Singapore department of geography and now on extended leave from the University of British Columbia, has been appointed Director of the Institute of Southeast Asian Studies in Singapore.

This relatively small group of local geographers already has a formidable list of publications to its credit. Tunku Shamsul Bahrin has written at least six substantial papers, mainly on the Indonesian migration to Malaya and on land development and settlement in Malaysia, Indonesia, and the Philippines,[69] while Dr. Hamzah Sendut has a similarly numerous total, concerned primarily with rural settlement and resettlement and with urbanization in Malaysia.[70]

Of Professor Ho's eleven main papers more than half of them written after he had moved to Canberra in the mid-1960's, the majority deal with agriculture and/or rural settlement in Malaya but his earlier interests in physical geography find expression in two excellent short studies of the Malaysian environment.[71] With six of his eight main Malayan papers dealing with agricultural settlement Professor Wikkramatileke clearly had much in common with Professor Ho, though he has also produced two very good general studies on Singapore and in three papers written in collaboration with other scholars, he has extended his range still further.[72]

Professor Ooi likewise has focussed regionally upon Malaya and Singapore, in his shorter works and in his extended study of "Mining landscapes of Kinta" (1955), his monograph on "Rural development in tropical areas with special reference to Malaya" (1959), and most obviously in his book Land, People and

---

[69] See, for example, his "Land Conflicts in the Tanay Resettlement Project (Rizal), Philippines," Journal of Tropical Geography, Vol. 27 (1968), pp. 50-58, and "A Preliminary Study of Fringe Alientation Schemes in West Malaysia," ibid., Vol. 28 (1969), pp. 75-83.

[70] See, for example, his "The Resettlement Villages in Malaya," Geography, Vol. 47 (1962), pp. 41-46, and "Patterns of Urbanization in Malaya," Journal of Tropical Geography, Vol. 16 (1962), pp. 114-130.

[71] Robert Ho, Environment, Man, and Development in Malaya, Inaugural Lecture, University of Malaya (Kuala Lumpur, 1962); "The Environment," in Wang Gungwa (ed.), Malaysia--A Survey (London, 1964), pp. 25-43. The other works mentioned appear mainly in the Journal of Tropical Geography (1962, 1965, 1969).

[72] See, for example, his papers in the Journal of Tropical Geography (1958, 1964), Pacific Viewpoint (1962, 1964, 1972), and his recent study with Karpal Singh of "Tradition and Change in an Indian Dairying Community in Singapore," Annals of the Association of American Geographers, Vol. 60 (1970), pp. 717-742.

Economy in Malaya (1963), which is recognized as the standard work on the subject. Moreover, he has recently contributed to, and acted as joint editor, with Chiang Hai Ding, of Modern Singapore (1969), which was produced to mark the 150th anniversary of the founding of that city.

So far as Dr. Sandhu's several publications are concerned, settlement within Malaya is once again the central theme, but he more than any of his colleagues has focussed his attention upon the Chinese and Indian communities, and in his Indians in Malaya--Immigration and Settlement 1786-1857 (1969), has produced a study which has been widely acclaimed by geographers and historians alike. Lastly, K. E. Chan, L. S. Chia, K. S. Loh, T. F. Lam, P. K. Voon, P. P. Wong, Y. M. Yeung, and other more recent staff members of the geography departments of the Universities of Malaya or Singapore have each published one or more research articles on some aspect(s) of the local geography.

Besides the various institutions in Malaya and Singapore already mentioned above, Nanyang University was founded in Singapore in 1956 to cater primarily for students from Chinese-language schools and more recently the Universiti Kebangsan Malaysia (National University of Malaysia) was established at Kuala Lumpur with an analogous emphasis on Malay language education. Both of these universities contain geography departments, that at Nanyang having been initially staffed largely by scholars originally from mainland China, Taiwan, or Hong Kong, while the Universiti Kebangsan has recruited mainly from the University of Malaya and from Indonesia.

Among Nanyang University publications, History and Geography (1962) contains an article (in Chinese) on the development of the port of Singapore by Tsou Pao-chun, formerly in charge of geography there who has also published briefly in English on both Singapore and Kuala Lumpur. More recently Nanyang has published Geographical Papers (1972), which, inter alia, includes two in English, respectively by the present Chairman, Dr. Ang Koh Ping on agricultural development in Singapore, and by Tan Kok Chiang on the international rice trade since the Second World War, as well as others in Chinese, on the potential for agricultural development in western Sabah, by Niew Shong Tong, postwar urban growth in Southeast Asia, by Chiang Tao Chang, the trade between Japan and Southwast Asia by Phua Beng Tee, and the changing patterns of entrepot trade in Singapore by Yen Sen Shun. Meanwhile Universiti Kebangsan Malaysia has inaugurated an impressive new interdisciplinary journal, Nusantara, published partly in Malay and partly in English, the second issue of which contains a geomorphological article "Sediments of the Kelantan delta: grain size and distribution in different environmental conditions" (1972), by one of its editors, Zakaria Hj Awang Soh, who is Acting Head of the Department of Geography.

By comparison with the great productivity of Malaya and Singapore, the meagre output of geographical publication in Indonesia may suggest that the latter is a geographical backwater. This is very far from the truth, though the task of building up academic geography virtually from scratch during the troubled years following the achievement of independence was beset with immense difficulties and much still remains to be done. Nevertheless, with the help at first of F. J. Ormeling and H. Th. Verstippen, and later of J. N. Bhatta, the Indonesian Army's Geographical Institute, Balai Geografi, has been strengthened and its publications include four items by Bhatta himself on various aspects of population studies. More recently Kol. Dr. I. Made Sandy has organized the production of a national series of land-use maps, many of which are now available, as also is his useful economic atlas of Indonesia (1971).

Within the Indonesian universities geography is concentrated primarily in Gadjah Mada University, Yogyakarta, whose Fakultas Geografi now has the largest staff of any geography department in Southeast Asia. During the critical period in the late fifties and early sixties its development was immeasurably assisted by the devoted service of M. Halim Khan (M.A. Aligarh, Ph.D. Clark) who, besides playing a major role in teaching and in training junior staff, also succeeded in launching Madjalah Geografi Indonesia: Indonesian Journal of Geography) to which moreover he contributed numerous articles mostly based on his pioneer local research. In this activity he was joined by several of his Indonesian colleagues, among whom Karmono Darmojuwono (geomorphology), R. Bintarto (human geography), and Basuki Sudihardjo (cartography) remain as senior members of an increasingly diversified and talented staff. Although publication of this journal has been somewhat erratic, it has played a vital part in promoting local research and in establishing geography in Indonesia.

Nevertheless only a small fraction of the research done by Indonesian geographers has been published in this or any other journal, for much the greater part of their contribution has been directed towards urgent practical tasks, most obviously in various aspects of planning, and in addition the senior scholars are heavily committed to the training of more geographers for whom the demand is great.

Outside Gadjah Mada some geographical training is available in the University of Indonesia at Jakarta, under Kol. Azwar Hamid and others, and also in the Department of Regional and Urban Planning of the Institut Teknologi

Bandung. Although only one of the latter's staff, Djohara Djajadinata, was trained as a geographer, others have in effect become so through practising the craft. Among these W. J. Warworoentoe is outstanding, and his several exceptionally interesting mimeographed papers, some in English and more in Indonesian, deserve to be more readily accessible. Meanwhile, it is encouraging to be able to report the publication of the first book by an Indonesian, writing in English, on the geography of his own country, namely J. Hardjono's Indonesia--Land and People (1971).

The development of geography in the Philippines since the achievement of independence in 1946 shows certain similarities with that in Indonesia. Thus here also the basic task arose from the need to make up for the neglect of the subject under the colonial regime, but new local initiatives in this direction were similarly helped by individual scholars from the former metropolitan power. In this process a key development was the creation of the Philippine Geographical Society in December 1950, thanks largely to the efforts of the late Dr. J. M. Feliciano, Head of the combined Geology and Geography Department of the University of the Philippines, the irrigation expert Dr. Dominador Z. Rosell, and Dr. Alden Cutshall, then serving as a Fulbright Professor of Geography in Manila. Moreover this new society proceeded to establish its own publication, The Philippine Geographical Journal in 1953 which, despite a serious gap between 1958 and 1963, has since become firmly established.

From the outset the Philippine Geographical Journal received invaluable support from American geographers interested in the Philippines, many of whom contributed articles to it, and although the number of professional geographers in the country remains minute, Dr. Telesforo W. Luna of the University of the Philippines has made a notable contribution in his articles on regional geography and in particular with his long paper "Manufacturing in Greater Manila," 1964. Likewise Dr. Rosell, who has become a de facto geographer, has contributed extensively within his special field of water resources and irrigation and also on the geographical approach to Philippine economic development. Meanwhile many other Filipinos prominent in various fields, and likewise appreciating the importance of geography to national needs, have written for the journal on local applications of their own expertise. These include P. Baldoria on political geography, S. R. Cruz and T. M. Mandoza, agricultural meteorology, A. M. Dalisay, rural communities, J. Gutierrez, agricultural land use, W. A. Minosa, earthquake problems, and T. S. Oracion, kaingin (swidden) cultivation, while E. B. Manalo has contributed what amounts to a monograph on the distribution of rainfall within the Philippines. In addition the journal has published several brief accounts of particular parts of the Philippines by people with appropriate local knowledge and numerous articles covering virtually every commodity produced in the country, even including cobra venom.

All of this adds up to a rich store of published data useful to geographers, and may perhaps be regarded as the local counterpart, within a very differently organized system, to some of the work carried out under government auspices by applied geographers in Indonesia. But it is probably the absence of any significant governmental demand for geographers which more than anything else holds back the development of the subject in the Philippines today.

In Thailand finally one finds another variation on this same theme. Here, in the absence of any Western colonial contact, the national university system reflects a variety of Western traditions but with a somewhat uneven time-lag. As a result geography has hitherto been accorded the kind of minor role which it had in Britain before the First World War, being taught until very recently only as a junior partner of history at Chulalongkorn University, Bangkok, and accordingly it has not developed to the stage of generating any significant research.

However, the activities of many Western and Japanese researchers in their country have recently alerted the Thai authorities to the practical contribution which geography can make to development, and in the last few years new posts have been created in geography at Kasetsart, Thammasart, and Ram Kamhaeng University, Bangkok, Tap Kaen University, Nakorn Patom, and the University of Chiengmai, and there is also a particularly lively and expanding Department of Geography in the College of Education at Prasanmit, Bangkok. As all of the teaching posts have been filled by young Thai graduates, most of whom have either had or will shortly receive further training overseas, the end would seem to be in sight of the present anomalous situation in which geographical research in the only Southeast Asian country which has never been under colonial rule is virtually monopolized by expatriates.

The Outlook Ahead

Following the short but intellectually stimulating period when the Pacific War led to a new awareness of Southeast Asian geography, the last quarter century has seen a curious reversal in certain respects. Before the war, geographical knowledge of Southeast Asia was extremely patchy, with the fullest detailed coverage devoted to Vietnam and Burma, very much in that order. Today the overall picture is altogether fuller, but Vietnam, as a casualty of war, and

Burma as a casualty of introversion have come to represent the most serious regional cases of contemporary geographical ignorance.[73] While prophecy in such matters is always risky there have been recent signs of movement in Burma, and there is now a possibility of more significant change in Vietnam.

Meanwhile Malaysia and Singapore which had only just begun to receive serious geographical attention on the eve of the War have since been intensively studied from the standpoint of every major aspect of geography, and indeed are now more effectively covered in this respect than all but a handful of the so-called developed countries. The Philippines also is by now well served in terms of contemporary social, and economic geographical analysis,[74] and a beginning has been made by D. F. Doeppers and others in studying its historical geography, though there is still a need for more detailed regional work, particularly on problem areas.[75] In Thailand, on the other hand, a relatively small number of Western and Japanese geographers have made important specialized studies in both economic geography and applied geomorphology, but no attempt has yet been made to produce a really up-to-date general account of the country's geography which in many respects has been profoundly changed since Credner's time. The essence of these remarks about Thailand also applies with even greater force to Indonesia, with the major qualification that no basic geographical study comparable to Credner's has ever been made of Indonesia as a whole.

Clearly, therefore, there is now need for something approximating to definitive geographical studies of several of the individual Southeast Asian countries. But if such studies are to be attempted it must be hoped that the authors concerned will try to see these countries in terms of the ongoing process of human-ecological adjustment and not merely as new areas in which to try out the latest fads of the so-called new geography. That, however, is in no way to deny the urgent need for more geographers to interest themselves in develop-

---

[73] Presumably much useful but at present classified material now exists on both halves of Vietnam and, at least in part, may be reworked for publication. Unfortunately, I am not familiar with current geographical research in either the U.S.S.R. or China but it would be surprising if it did not contain some useful material on Vietnam.

[74] A new book by T. M. Burley, The Philippines--A Social and Economic Geography, appeared early in 1973.

[75] K. Hausherr is a pioneer contributor in this respect. See his Die Entwicklung der Kulturlandschaft in den Lanao-Provinzen auf Mindanao (Philippinen) unter Bericksichtigung des Kulturkontaktes zwischen Islam and Christentum (Bonn, 1972).

mental studies and to play an active part in planning, and it is obviously desirable to have more case studies of the developmental process in individual Southeast Asian countries along the lines pioneered by Missen's Viewpoint on Indonesia.

Nevertheless I am convinced that this kind of work, important though it is, should not be allowed to dominate the scene, for even applied geography may itself go dangerously astray if pure geographical study is neglected. Here there is unquestionably an enormous amount awaiting attention in the geomorphology and biogeography of virtually every country in the region. But perhaps of even greater long term importance is the need for more historical depth in our studies of all aspects of the human geography of Southeast Asia though, as has already been implied, Wheatley, Jackson, and Sandhu have made an impressive contribution in this connexion in relation to Malaya, and Doeppers, Cobban, and Jackson have followed their lead with respect of the Philippines and Indonesia.

Even so, we know very little as yet about the ecological significance and consequences of the great era of Indian acculturation, and are similarly ignorant about the corresponding effects of the diffusion of Islam. And although rather more information is available about the contribution of the immigrant Chinese to the modern geographical transformation of Southeast Asia, no one has yet done anything like justice to this theme from a geographical point of view. Compared with the immense literature stemming from the experience of European settlement in the New World, published research into the in many ways more geographically significant migration of southern Chinese from a warm temperate environment (but with a culture developed primarily in cooler latitudes) to the humid tropical and equatorial zones of South-east Asia is surprisingly meagre. If, as has been claimed, Dr. Joseph Needham's work on Chinese science has transformed the history of science, such a study as I have suggested might lead to major changes in geographical thought.

In the same general direction we need to know far more about the truly astonishing population growth which has taken place in Southeast Asia since the early nineteenth century. Almost certainly its causes were far more complex, and the variations from place to place much greater than anyone has yet demonstrated, and undoubtedly geographers ought to be able to provide useful insights here.

But above all, I would suggest, the fact that a new generation of Southeast Asian geographers is now coming to the fore should add a new dimension to our work. Just as the great Swiss historian Burckhardt saw history as the study of what one period finds significant in another, so I believe geography should be the study of what the people of one place find significant in another. With both Westerners and Southeast Asians working side by side we may now at last aspire to something like synoptic vision, from which alone a true appreciation of depth can come.

Such comments as these may perhaps seem old fashioned to some of what I call our new[2] geographers, but to be out of fashion is not necessarily to be out of data and the kind of problems I have mentioned are the sort that our non-geographical colleagues in area studies want answered and increasingly they are turning to us for assistance. Already geography has contributed much to area studies and along such lines it can give still more. And in co-operating with other colleagues in area studies, geographers will also benefit themselves, for they will in the process be led back to the heart of their own subject. For we too have tended in recent years to float in a limbo of disembodied space. We need to remind ourselves that geography is not the study of abstract space; it is the study of places, real, earthy, inhabited, and alive.

CHAPTER 10

GEOGRAPHY AND AREA STUDIES: THE CASE OF JAPAN

Norton Ginsburg

University of Chicago and Center for the
Study of Democratic Institutions

For decades the Japanese have been describing themselves as a "small, poor, island country," and most of the geographical world believed them. Of course, many Japanese also believed in it, and the more ambitious of them, as members of an authoritarian ruling elite, led their country into adventures which led, or so it appeared for a time, to disaster. Yet, out of the ashes of smouldering cities and despite the loss of most of the modernized industry of the country and millions of the flower of Japanese manhood, Japan has risen like a phoenix to the point where it could be described by Herman, the great Kahn, himself, as potentially the most powerful country in the world.[1] How could this self-perception have been wrong, or was it there in the first place? How could geographers with their tradition of a territorial _gestalt_ composed of distributional, ecological, and historical dimensions fail to perceive what really was the case, that wealth and poverty are not functions of natural resource endowments, or at least endowments alone, but rather of complex and mysterious synergetic relationships which characterize the complicated systems we call societies.

In 1935 Japan was recognized to be a comer by most observers, although many still regarded it as merely quaint. After all, it had gained a kind of internationally recognized inferiority through the Washington Conference agreements of 1921-22; and it had recently occupied Manchuria and tweaked the noses of the League of Nations in doing so. It controlled Korea, but not many knew just what that might mean since few Americans other than the McCune brothers had much understanding of what was involved; and Taiwan was virtually _terra_

---

[1]Herman Kahn, _The Emerging Japanese Super State: Challenge and Proposal_ (Englewood Cliffs: Prentice-Hall, 1970).

incognita to other than its residents and the Japanese. On the other hand, not only these foreign relations and enterprises were cloaked in uncertainty and imprecision of information. Japan itself was almost as "closed" as a black box, closed among other things by an extraordinary language which was frustrating to all foreigners and even to some degree to the Japanese themselves, and by a remarkable social organization and complex history which was only beginning to be studied by a handful of foreigners. Thus, in 1935 in the United States, only rather few academics (reportedly 111) were capable of interpreting at least some of the aspects of Japanese society, economy, and polity. Most of these were historians, and only twenty-one, eight of whom were native Japanese, were able to use Japanese-language sources in their research.[2] In spite of the fact that the Japanese even then had virtually universal literacy and ranked high among the countries of the world in books and journals published, it was virtually impossible to study the Japanese language in the American higher educational system with the exception of several institutions, and even then primarily on a tutorial basis. What was going on in American geography at the time?

In fact, very little, when numbers are taken into account. It is often thought that in some Golden Age before the "Barbarians" destroyed the "civilization that was Geography" some twenty years ago, geographers devoted themselves to studying and seeking understanding of other lands and peoples, as well as their own. In fact, with some noteworthy exceptions, there was no such Age and precious little such activity. Regional geography, defined largely as the study of foreign areas, was as limited, in absolute terms even more so, than it now is. There was not, for example, more than one American geographer who had any substantial interest in India--Wellington Jones--and he also was one of the three or four who shared a major interest in Japan as well.

None of the men interested in Japan had a working knowledge of the Japanese language and therefore access to the considerable literature, particularly in historical geography, that the Japanese had been creating. Of course, they could use collaborators and interpreters, as did Trewartha and Hall for example, and perhaps this was as much as could be done under the circum-

---

[2]See John W. Hall et al., "Japanese Studies in the United States: A Report on the State of the Field, Present Resources, and Future Needs," reprinted in part in The Newsletter, Association for Asian Studies, September, 1971, Supplement No. 1 (1971), pp. 8-9. This report was prepared by the SSRC-ACLS Joint Committee on Japanese Studies, no member of which was a professional geographer, of which more below.

stances, but it placed severe constraints on what could be done and how. It is interesting to note that Jones, for example, never put his students to work on mapping isopleths of crop ratios and land uses in Japan, as he did so admirably in India.[3] The reason was that the Japanese sources were hard to come by in the first place, and most of them were in Japanese (even to table headings in some official publications) on the other.

Yet, the handful of men involved did yeoman service. In spite of their limited period in the field, their lack of a language competence, and limited scholarly resources, they created a modest literature and trained students in sometimes remarkable ways. Jones in particular looked at what he called "the human occupance of regions" as complex systems of relationships between man and nature, with culture intervening, no doubt owing much to Sauer's influence when the latter was at Chicago. He knew much about nature, less about man, and still less about culture, in Japan at least, but his focus of attention was sound. Moreover, it encompassed that other major thrust of geographic effort, concern with the "organization of area," a concept associated also with the name of Robert Platt and with what came to be regarded as the "Chicago school" of geography. Several Japanese geographers had been working on the patterns of settlement distribution and spacing, using quantitative techniques, such as I. Matsui, and he corresponded with them.

The value of large-scale topographical maps also was a hallmark of Jones' approach.[4] Relatively few of these on Japan proper were available in this coun-

---

[3]See the admirable series of Masters theses on Indian agricultural regionalization done by his students at the University of Chicago in the 30's.

[4]One of the advantages of doing research on Japan is the availability of complete and recent coverage of the country on topographic quadrangles. The standard scale is 1:50, 000, and sheets are corrected at approximately five-year intervals. These maps are enormously detailed and contain a wealth of land-use information not available on their U.S.G.S. equivalents in the United States. Larger scale coverage is available for metropolitan areas. In addition, a large proportion of the country, perhaps 60 per cent, is now covered on land-use maps which use the 1:50, 000 topographic quadrangles as a base. These were done under prefectural auspices and therefore are not as readily available as the topographic quadrangles, which are produced by a national agency. All maps, of course, are in Japanese. However, during the Pacific War, English-language, multi-colored versions of the topographic sheets were issued by the Army Map Service and related agencies. When available, these may be used both for purposes of examining sequent occupance and to overcome at least part of the language barrier. The value of this great resource in studying the regional human occupance of Japan, particularly in rural areas, is enormous. Its value for teaching purposes is almost as great. In addition, aerial photo coverage is available for selected portions of the country on a commercial (and therefore

try, and the language barrier to using them was considerable, but they were recognized as an essential basis for landscape study, as a point of entry into area analysis, and as vast storehouses of information, the retrieval of which was difficult but rewarding. In his work the word "space" either in its noun or adjectival forms was used very little, but the word "area" in both of those forms was used a lot. Whereas "space" refers to a three-dimensional void and therefore was of little interest, "area" was pattern, content, and organization, all rolled into one.[5]

It is difficult not to digress, it appears. The fact is that geographical knowledge in the United States about Japan in the thirties was highly limited, if in itself not unsound, and it was held pretty much by three or four men who had only indirect access to the Japanese literature and only limited periods of field work in Japan itself. There is an old Japanese saying which translates (badly) something like this: "If you ask me why, I can only reply: I don't know." That is the case here, except for the observation that there were few parts of the non-Western world that could be described differently, at least insofar as geographical knowledge and its American practitioners were concerned. India already has been alluded to; most of Africa surely also illustrated the case; Latin America was a partial exception. Geography in America, even then, was highly parochial, culture-bound, and ethnocentric.

One concluding point should be made, however. Given the very small population of Japan specialists in this country at that time, the fact that there were less than a handful of geographers with that specialty suggests a better record than one might suppose. Moreover, a number of useful publications on Japan appeared in the geographical periodical literature which were highly useful and

---

expensive) basis. The utilization of these materials as the basis for a geographical data storage and retrieval system has not yet been explored, although precedents exist in India and elsewhere.

[5]See my article on "Area" in the International Encyclopedia of the Social Sciences, Vol. 1, pp. 398-401. As Jean-Francois Revel has noted in conversation, vocabularies evolve as a response to a felt need for breaking with convention, for pushing an "intellectual fashion," and often the meaning of words is perverted in the process. Thus, in the current fashionable jargon, "space" has come to mean "place," "content," and "organization," which conflicts with common usage of the word. Thus, when E. A. G. Johnson (not a geographer) entitles his culture-bound volume, The Organization of Space in Developing Countries (Cambridge: Harvard University Press, 1970), he falls plumb into this semantic trap.

even exemplary,[6] and Trewartha's Reconnaissance Geography of Japan, a major work done with the help of Japanese collaborators it is true, was a model regional text of its day, largely descriptive but analytical as well.[7]

The Current Situation

In spite of the alleged "progress" in geographical studies since the second World War and the substantial increase in the practitioners of the geographers' art, the present situation is in many respects no better than it was thirty years ago and in some respects worse. Of the 416 persons estimated to be authentic Japan specialists and who make up the field of "Japanese studies" as defined by the Joint Committee on Japanese Studies, only eight are geographers. In fact, the figures need to be qualified. They do not include persons in Canadian universities, for example; nor do they include all persons who responded to a recent questionnaire distributed by the Association for Asian Studies about area specialization, since 717 respondents listed Japan as one of their areas of interest. Moreover, many individuals who have shifted their major career lines from scholarship to, say, journalism or administration also are not included, and geographers proportionally have a higher attrition rate in these directions. Nevertheless, as a rough approximation, the data are illuminating and rather discouraging, given one of the presumed major missions of geographic enterprise--to increase understanding of the global ecosystem, with both cultural and

---

[6]For example, R. H. Hall, "Some Rural Settlement Forms in Japan," Geographical Review, Vol. 21 (1931), pp. 93-123, and idem, "Tōkaidō: Road and Region," Geographical Review, Vol. 27 (1937), pp. 353-377.

[7]Glenn T. Trewartha, A Reconnaissance Geography of Japan (Madison: University of Wisconsin Press, 1934). This volume was the basis for his highly useful Japan: A Physical, Cultural, and Regional Geography (Madison: University of Wisconsin Press, 1945), and it, in turn, was revised as his Japan: A Geography (Madison: University of Wisconsin Press, 1965), still the most useful reference work of its kind on Japan, even though it is locked into a regional format based in part on physiography. This drawback (in my view) is mitigated in part by the wide acceptance in Japan of broad regional subdivisions of the country that conform roughly to that format, e.g., Tōhoku and Kantō. Thus, the problem of breaking out of an obsolete physical regional shell is not as acute as it has been in India, for example. When about fifteen years ago I proposed to W. Norman Brown that the "regionalization problem" for South Asia could only be solved intellectually through the determination of a regional system based upon the four great Indian metropolises, it came to him as an extraordinarily original proposal. In fact, of course, it was not. The Natural Resource Planning Board had tried to do the same thing in this country, under the influence of several thoughtful geographers, nearly thirty years ago.

TABLE 1--JAPAN SPECIALISTS IN THE UNITED STATES

History	110
Language-Linguistics	98
Political Science	52
Literature	33
Anthropology	28
Economics	22
Art-Architecture	16
Sociology	15
Philosophy-Religion	15
Geography	8
Education	5
Psychology	4
Other (music, drama, etc.)	9

physical elements, and to identify and comprehend the national and regional subsystems that in sum constitute the global system.

Of course, historians account for the largest single group, then those in language-linguistics, as Table 1 shows. Geographers are well toward the bottom of the list. Their performance in absolute terms perhaps is not so poor as it might seem. There are only fifteen sociologists, for example, listed, and there are more than three times as many sociologists in the country as there are geographers. However, the anthropologists, who number about the same as the geographers as a body, have more than three times as many specialists on Japan, and even the economists rank considerably higher.

The situation is about the same when one looks at publications on Japan, as exemplified by doctoral dissertations completed since 1947. According to the rather narrow definition of what is acceptable to the Joint Committee on Japanese Studies, 348 dissertations were completed in the Japan field between 1947 and roughly 1969. Of these, as indicated on Table 2, about one-third were in history, but only five were in Geography, presumably one of them mine. Again, there must be some qualification because these studies were restricted to those concerned entirely with Japan and at least in part dependent upon

TABLE 2--DOCTORAL DISSERTATIONS ON JAPAN, 1947-69

History	112
Political Science	54
Economics	38
Literature	34
Languages and Linguistics	25
Anthropology	19
Religion	17
Sociology	16
Art	10
Geography	5
Education	4
Psychology	1

Japanese sources. The number completed on a looser basis of classification would be 1,103. In any event, Geography's contributions rank low on the scale. It is particularly significant that Economics ranks as high as it does, since it is the social science least disposed to favor areal specialization. What, then, does all this suggest?

First, it suggests that in general area specialization in American geography is less than it ought to be, if Geography as a discipline were to fulfill its multifaceted mission. It suggests also that there may be difficulties in working in foreign areas which have special relevance to geographical enterprise. Third, it implies that Japan itself may be a particularly difficult area upon which to specialize, especially because of the language problem, since accessibility is not, as it is in the case of China Studies, a major issue, and the amount of material available is truly enormous, although it is not always in readily available form.

Of course, it is true that specialization on Japan presents great difficulties both in training and research, but particularly in the former. Apart from anything else, a reasonable command of the language requires at least three and usually more years of intensive study, and only part of this training can be acquired during the normal programs leading to the doctorate in Geography, as well as in other fields. The commitment involved, then, must come early, and it requires a high degree of dedication which is hard to expect in the current context of geographical studies in America. So long as it is believed that one can learn as much by studying one's own backyard, so to speak, as by studying a foreign area, and so long as behaviour and institutions, and therefore distributional patterns and areal organization, are asserted to be independent of culture (improbable though that might appear), then one can hardly expect students to decide on a career line that will seem to disadvantage them relative to their peers in the competition for jobs that has reached a recent peak in the profession.

Equally important is that fact that the Japanese themselves are major producers of a geographical literature. Several geographical periodicals are published regularly in Japan, at least three of them for a long period of time; and they are of high quality. The Japanese read everything, and they select from the literature those concepts and techniques that appear to them to be most useful. One cannot work on Japan without knowing what the Japanese themselves are doing, and that requires the ability to read Japanese. To some extent, this problem is mitigated by the recent increase in the publication in English of some valuable geographical studies, but these account as yet for only a small proportion of the total literature.[8] Certainly, intensive field work on the micro- or behavioural level cannot be carried on without access to Japanese-language sources and without an ability to handle the spoken language at least to some degree.

Nevertheless, the fact remains that there are relatively few institutions in this country which provide opportunities for proper training in Japanese geographical studies. These would include, in addition to this University, less than a dozen in all which would have strong programs both in Geography and in Japanese studies.

In spite of these problems, geographers know more about Japan than ever before. In part this is because of the contributions of the Japanese geographers themselves. In part it is because all of the small number of Japan specialists in Geography have been thoroughly trained in their own fields on the one hand and their sensibilities have been heightened by their familiarity with Japanese culture on the other. Thus, they tend to look at landscape and spatial processes alike with a hybrid perspective that, like the hybrid strains of certain grains, gives larger returns, in this case in the form of deeper understandings.

---

[8] For example, the well-known Science Reports of the Tōhoku University, 7th Series (Geography), which has been appearing since the early 1950's; and the Association of Japanese Geographers, _Japanese Cities: A Geographical Approach_, Special Publication No. 2 (Tokyo: 1970), edited by Shinzō Kiuchi.

The insights involved are difficult to describe and perhaps impossible to quantify, but that does not make them insignificant. To the observer trained within two cultural contexts so to speak, Osaka looks not so much like a slightly exotic equivalent of Chicago, as a distinctive element in a complex landscape pattern, resulting from "fuzzy sets" of circumstances, some indigenous, some foreign, which is characteristically Japanese. Thus, when Douglas Eyre examined the operation of a rural water control system in Okayama Prefecture, he did so as much in terms of the socioeconomic matric within which it is set as the engineering and agronomic problems it was intended to solve.[9] Similarly, the Japan discussed in David Kornhauser's forthcoming book is presented as a cultural ecosystem with its state variables, particularly urbanization, described in ways that would not have been possible thirty years ago.[10] Conversely, familiarity with Japanese cities may make it possible to see ours in fresh perspective.

Japanese Area Studies

The second World War provided an important fillip to the expansion of area studies programs in general. The need for area specialists was demonstrated as soon as the nature of the conflict became global; and the American intelligence establishment quickly sought out those individuals who had some special familiarity with given countries and cultures. For the most part, that establishment came to be organized on area lines. Historians predominated, as might have been expected, but geographers assumed a much larger role than their relatively small numbers would have suggested. The ability of geographers to deal with a heterogeneity of information about countries and to put it into some readily comprehensible form was recognized early on. Clearly, there was something in the geographical perspective on area and polity that was distinctive. In his book Strategic Intelligence, which defined the ideal-typical model of such an organization, Sherman Kent recognized, at least implicitly, this capability and potentiality.[11]

---

[9] J. D. Eyre, "Water Controls in a Japanese Irrigation System," Geographical Review, Vol. 45 (1955), pp. 197-216, and his "Japanese Land Development in Kojima Bay," Economic Geography, Vol. 32 (1956), pp. 58-74.

[10] David Kornhauser, Japan: Transformation and Tradition (tentative title) (Chicago and London: Aldine and Longmans, in press).

[11] Sherman Kent, Strategic Intelligence (Princeton: Princeton University Press, 1949), and my review of it in the Annals of the Association of American Geographers, Vol. 40 (1950), pp. 89-91.

As a result, interest in area studies in American universities rose.[12] One of the first major post-war reports on area studies, especially as related to the social sciences, was prepared by Robert B. Hall for the Social Science Research Council, of which he served for a time as Vice President.[13] Hall, of course, also is a geographer, one of the few whose major research interests had focussed on Japan. In his report, he explored most of the problems that now arise with regard to the rationale for and organization of area studies and their relations with the social science disciplines--the language-training program, field work, collaborative research among the disciplines, the need for disciplinary as well as culture-area training, and the relations of area training to degree programs in American universities.

Hall himself was largely instrumental in establishing in 1950 the famous field station in Okayama of the University of Michigan Center for Japanese Studies. This institution was perhaps the most influential single enterprise in the general field of Japanese area studies, as well as in Geography. Out of it came many of the key names associated with contemporary Japanese studies, as well as several historians. So far as can be ascertained, about one out of four of the Japan specialists in this country over the age of 35 owe much of their continued interest and a high proportion of their training experience to the Center. Several geographers are included within this number. Moreover, the effects of interdisciplinary training and research experience through a field station placed its mark on the ways in which scholars, as they matured and carried on their research, perceived their problems.

From the standpoint of geography, one of the major effects was introduction of a geographical perspective into the professional writings of such leading Japan studies figures as John Hall, Richard Beardsley, and Robert Ward, among others. The closing of the field station in 1955 was inevitable, of course, given the financial stringencies that always accompany activities presumably marginal

---

[12] For a discussion of these and related developments, see Bryce Wood, "Area Studies," in the International Encyclopedia of the Social Sciences, Vol. 1 (New York: Macmillan, 1968), pp. 401-407.

[13] R. B. Hall, Area Studies: With Special Reference to Their Implications for Research in the Social Sciences (Pamphlet 3) (New York: Social Science Research Council, 1947). About the same time there appeared William Fenton's, Area Studies in American Universities: For the Commission on Implications of Armed Services Educational Programs (Washington: American Council on Education, 1947), and the following year Charles Wagley prepared Area Research and Training: A Conference Report on the Study of World Areas (Pamphlet 6) (New York: SSRC, 1948).

to the main interests of the American university; but, for a time, it might also be said to have served its purposes. Moreover, it was primarily concerned with regional and rural problems rather than urban ones which were coming to loom so large. Nonetheless, if one were to ask today what sort of institution might well be lacking from the array of resources available to the incipient specialist in Japanese studies, an urban-oriented field station in Japan, run much on the lines of the old Okayama field station, might well rank high on the list of priorities.

In any event, for a time, Geography did quite well in the burgeoning of area studies programs in the United States after 1947. As reported in W. C. Bennett's Area Studies in American Universities,[14] only a few of the 29 major programs on various major regions of the world did not have geographers present in their staffs, but the paucity of area-trained geographers was noted even so.

In time, of course, the role of area studies in higher education came under even more intensive scrutiny. Most of the problems raised by Hall in his initial study had not been satisfactorily solved. The highly specialized disciplinary orientation of American higher education, which had been becoming more flexible for a time under the impact, first, of the Depression, and then, the War, strongly increased. In spite of substantial government and Ford Foundation aid to such programs, they came under ever stronger pressures making for fragmentation. Even though many have retained their identity and are doing yeoman service, in fact their influence both nationally and within individual institutions has declined, and most may be said to provide merely an umbrella for the essentially unrelated activities and interests of a congeries of scholars rather than acting as a collegium. Nonetheless, the rationale for area studies continues appealing to many and strong to some, certainly to most humanists.

The Case for Japanese Studies

The arguments for area specialization on Japan have been well presented by the SSRC-ACLS Joint Committee report referred to previously:[15]

> The case for Japanese studies rests upon the firmest of premises: that the United States and Japan have common interests of the profoundest sort, that our relations with Japan in most fields will be as important and often as

---

[14] Area Studies in American Universities (New York: SSRC, 1951).

[15] J. W. Hall et al., op. cit., pp. 1-4.

close as with any other nation in the world, and that because of language barriers and differences in cultural background, there is a great need for conscious academic effort to develop understanding and communication with Japan.

But these facts are too often overlooked. Japan does not force itself upon the attention of the American public with dramatic arguments. It does not entice us with the grandeur of ancient civilizations; it has not been the mother of world religions; it does not pose overwhelming problems of population; it does not challenge us as an immediate threat. Yet the same is also true in large part of the other two major industrial complexes of the world--Western Europe, except for our cultural ties to its ancient civilizations, and the Soviet Union, except for its present threat potential. Actually, the Japanese historical experience is comparable in length and significance to that of France, England, or Germany; its feudal development--the only clear parallel to Western feudalism--is of particular comparative interest to those of us who share a feudal background. Among the great world religions, Buddhism and particularly its Zen form has been accessible to the West largely through Japan. And in modern times Japan's dramatic rise as a world power has had profound consequences for the United States.

Japan's modernization since the middle of the nineteenth century has been a noteworthy story involving moments of tragedy both for the Japanese and for us. Japan's surge forward in world standing in the past two decades has been even more remarkable. Japan now ranks third among the nations of the world in economic production. Its people have become major exporters of culture, science, technology, and scholarship. Its scientists and writers are known and respected in the West, just as its products are admired for their quality and inventiveness. Japan, in other words, has become a major part of that main stream of advanced international culture which was once monopolized by the West. There are those who go so far as to predict that Japan will dominate the twenty-first century. The prediction seems close to incredible to the layman, yet it rests upon informed predictions of the growth of Japan's gross national product. Japan's GNP, which in 1970 reached the $200 billion mark, is expected to double by 1975 and should again double by 1980, if the current rate of increase is maintained. This means that in 1975 Japan by itself will be producing as much as the combined economies of the members of the European Common Market today. Trade relations between our countries--the United States is Japan's first trading partner while Japan stands second only to Canada as America's most important trading partner--can only grow more extensive, complex, and delicate.

Our relationships with Japan are clearly as complex and crucial as with any other non-Western nation and, because of Japan's rapid growth, they will become proportionally more important to us in economic terms than Britain, France, or Germany and is likely to push to the fore in political, scientific, and cultural terms as well. Yet Japan does not hold anything like the same position in the American consciousness as the more familiar countries of Europe. The popular image is still that of an exotic and precocious but easily dismissed newcomer.

An even more enduring case can be made for the unique value of Japanese studies in intellectual terms--as an effective way to broaden the focus of American education beyond its traditionally narrow concentration on the Occident. Among the countries of Asia, Japan is the one which scholars can most directly and fruitfully communicate with their counterparts in the Western world. The development of the Japanese intellectual and academic community in the last century has opened up, as for no other Asian country, systematic knowledge about its history and society. The growth of Japan's "knowledge industry" has indeed been tremendous. There are today over 370 colleges and universities in Japan with a student population of over 1.4 million. Japan publishes annually over 40,000 titles in over half a billion copies, second only to the Soviet Union. Its scientists are responsible for an estimated one-sixth of all scientific publications in the world each year while translations from other languages keep the Japanese scholar abreast of intellectual developments throughout the rest of the world.

The fact that Japan has become so integral and important a part of the modern world community means that "the Japanese experience" has become directly relevant to the modern experience at large. Japanese data are as significant for American social scientists as those derived from Europe. Japan is the one Asian country in which theoretical or comparative studies can be carried out with the greatest prospect of success. In many areas Japanese scholars have broken new ground which we can ill afford to ignore. Increasingly any science or art which neglects the Japanese dimension becomes partial and parochial.

At present the dialogue between the intellectual and scholarly worlds of our two countries is remarkably active. But it is still only a fraction of what it should be and rests in unusually large degree on the effort of the Japanese to carry the discourse. For example, Japan translates nearly a thousand titles from English annually while we reciprocate with only 25 Japanese works in English translation. The American capability in the Japanese language is woefully inadequate. Our acquaintance with the work of Japanese scholars or the experience of Japanese men of affairs is for the most part filtered through inadequate conversion devices. There is no substitute for our own willingness to meet the Japanese half way in the effort to join our intellectual resources.

The economic, cultural, and scientific emergence of Japan thus poses a real challenge of understanding to the American people. The government needs more leaders capable of understanding the Japanese; so do private business and the mass media. The arts and sciences need more specialists who can convey the Japanese experience to American intellectual life. Our schools and universities need more teachers who can integrate an understanding and appreciation of Japan into the general education of our students. Our academic disciplines must acquire more ways to utilize Japanese data in their computations. Above all our people must learn to place Japan in a stable context of informed understanding--one not easily upset by political tensions.

Our national and scholarly interests obviously require the establishment of mutual relationships between our countries of the scale, closeness, and intensity that have been developed over the years between the American people and those of Europe. Yet we face unusual difficulties in establishing an effective community of understanding. Japan is the only truly world power which does not have a 'world language.' Its culture has evolved in comparative isolation for over two thousand years. In many ways Japan remains the most culturally distinct and intellectually inaccessible of the great powers, and this fact impedes greatly the exchange of ideas and sentiments with the rest of the world. To this extent Japanese studies require special attention and a continuing special support of the kind which has been given to priority areas of national concern in the past.

It is worth noting that the justification for Japanese studies is not predicated primarily on the importance of Japan in the contemporary world community, even though that argument can and should be made. A more powerful argument, especially as it relates to Geography and the other social sciences, is the extraordinary experience that marks Japan's modernization. For those of us concerned with problems of economic and political development and modernization, the case of Japan presents truly extraordinary appeal. It is clear that Japan, among all the non-Western societies, has "made it," and has done so largely on the basis of its own resources and most dramatically after having lost the greatest war in history, even if one spins off the Pacific conflict from the rest of the Second World War. There is scarcely any discussion of developmental problems that cannot be illuminated by reference to the Japanese experience, not only as it has been, but as it now is occurring and as it is becoming. Surely, geographers have something to offer along these lines, and I have discussed some of their possible contributions in my Presidential Address to the Association of American Geographers.[16]

In any event, what happened and is happening in Japan appears to have great relevance to what is going on elsewhere. That that development, however, defined, can only be examined in multi-disciplinary ways goes without saying. That a deep and comprehensive knowledge of a particular case, such as that of Japan, is a *sine qua non* for such an examination ought also to go without saying, but perhaps it is well to make the statement again.

The argument that Japanese studies provide an opportunity to break through the provincialism that characterizes many of the social sciences in America is perhaps of greatest significance, and it is an argument for area studies and for areal specialization in general. Moreover, it is widely recognized that Geography in America has become increasingly parochial and culture-bound and decreasingly relevant to extra-national and global issues. The Geographer has been likened to a plumber who carries a kit-bag of tools to any part of the world and applies them without regard for the complex social systemic settings within which they are to be used; and most of them after all have been based almost exclusively on Western experience, and highly limited experience at that. To take an example, the nature of Japanese urbanization (and for that matter elsewhere in most of the world) is scarcely understood, but in many of its basic parameters it seems to differ from American models. There should be nothing strange about this. We would not expect that Japanese family life, government, education,

---

[16] Norton Ginsburg, "From Colonialism to National Development: Geographical Perspectives on Patterns and Policies," *Annals of the Association of American Geographers*, Vol. 63 (1973), pp. 1-21.

interpersonal relations, perception of environment, and what have you, be the same as those in the United States, even discounting the great differences within this country. Yet, we may be told that cities are cities everywhere, that the same processes operate upon them, that cultural differences are of no importance. Fortunately, the plumbers of our profession have yet to find their way into the Japanese field, which is far too complex for them to be comfortable in, and Japanese Geographers themselves have been quite distrustful of concepts that bear little seeming application to their own societal environment.[17]

The ACLS-SSRC Joint Committee on Japanese Studies may be worth some mention in its own right, as its activities relate to Geography. No geographer is a member of the committee, but several of the members have had a vicarious relationships with Geography either through the University of Michigan Center and its Okayama field station, or through contact with the several geographers with special interests in Japan. Nor should this informal network of communication be under-estimated. There often is far greater empathy among area specialists than between individuals among them and other members of their chosen discipline. Even so, it is striking and significant that among the nine conferences in various fields relating to Japanese studies convened by the Committee prior to the issuance of its support, Geography was not one of them. On the other hand, the Conference on Modern Japan, which overlapped formation of the Joint Committee and which convened six seminars between 1961 and 1968 on problems of modern Japan, did have a geographer at one of its seminars, myself. It was concerned with modernization in Japan and brought together at Hakone about 25 social scientists and historians, half Japanese, half foreign, to discuss this crucial issue. With the discussions carried on in Japanese, there was some difficulty in communication from time to time, but the resulting volume, edited by John Hall, contained a number of propositions which were of distinctly geographical import.[18] Be all this as it may, the involvement of geographers in the Japan field clearly is not regarded as being of major significance by other Japan specialists, if only because of their small numbers.

---

[17] As illustrated, for example, by the articles by Morikawa and Watanabe in Kiuchi, op. cit.

[18] Perhaps I would have been invited to participate in other such seminars, but about that time, I shifted toward my China interests, and became a member of the Joint Commission on Contemporary China and eventually its Secretary for a two-year period. See Marius B. Jensen (ed.), Changing Japanese Attitudes toward Modernization (Princeton: Princeton University Press, 1965), especially the comments by J. W. Hall.

Moreover, although time has not permitted a systematic survey of the geographers who have applied for and received Foreign-Area Fellowships in the Japan field, a preliminary survey identified only one. Of course, geographers also have had access to their own program of foreign area fellowships at the pre-doctoral level administered by the Office of Naval Research for a number of years until 1969, but Japan was not an area to which students were attracted. The difficulty of the language, the foreignness of the cultural environment, the burden of piling area training upon the standard graduate programs, and the unfavorable environment evolving within the field itself, all have militated against the recruitment and the development of a solid group of younger Japan-oriented geographers.

Geographical Perspectives on Japan

It is said that: "It is more instructive to take a nap in Kyōtō than to study assiduously in any village." To a degree this aphorism can be applied to geography today. The "village" of contemporary geography in America at least may be associated with the minute slivers of interest, the innumerable trivia subjected to seemingly endless statistical manipulation, that seem to preoccupy so many of our colleagues. The "Kyōtō" of the saying is that complex set of ecological and spatial relationships which culminate in a country and society like Japan or a region like that of South Asia.

The extraordinary thing about Japan is that despite its size and complexity, it suggests the existence of a system of relationships in a way that the United States, for example, does not. The Japanese geographers seem to be aware of this, even though they might have been expected not to have seen the forest because of the trees. On the other hand, they have a keen sense of history, and what they see about them appears to be conditioned, understandably enough, by the evolutionary processes within their society. In the recent volume edited by Shinzo Kiuchi, one can perceive the sense of "wholeness" that permeates the work in urban and settlement geography in Japan. To the Japanese the concepts "urban" and "economic" are not to be equated. Rather, like the Greeks, "urban" and "civilization" are the key concepts.[19] The studies of Japanese geographers as exemplified in this volume, and in far greater number in the professional geographical publications in Japan, make it clear that urban life is not simply a

---

[19] Although the words themselves are of Latin origin, paradoxically.

matter of using shopping facilities, but is far more than that. Most Japanese geographers, in keeping with indigenous traditions of scholarship, are much concerned with the placement of particular events and activities within a broad social and geographical matrix. Many are deeply involved with questions relating to what might be termed a national land policy, both in retrospect and for the future.[20] This is striking contrast with current interests within the profession in the United States where an exhortation to deal with the problem of a national land policy, or even with resource management policies and matters of environmental conservation, would meet, on the whole, with responses of disinterest by many and contempt by some.

In fact, the recent Comprehensive National Plan for Japan reads as though it might have been written by geographers of remarkable sophistication and concern for the society of which they are a part. Although it is probable that direct geographical inputs were relatively small, the influence of men like Professor Kiuchi and his colleagues comes through with considerable clarity. The Plan looks at Japan as a one complex spatial and social system, and it urges that it be dealt with as such. Sectors in the conventional sense of national accounting and government planning are abandoned.

> The plan recognizes the fact, which geographers have long espoused, that environments are created by men in society, who mold them to their needs and wishes. It argues that environments must be created 'for humanity, and for . . . safety and comfort, as well as for efficiency.' It speaks of the need, even in Japan, for the development of 'indigenous, innovative technologies of our own,' and therefore by implication against the adoption of assumptions and premises which have their origin in the largely differing cultures of the West. It speaks of a new land-management system for national growth. It speaks of 'activity zones' within the national space-polity and space-economy, within which investment can be concentrated. Most important, it argues that Japan has now reached a point where, in what is called 'informational post-industrial society,' most places within the country can be made equally accessible. In effect, for the first time since the close of World War II, a national plan in a non-Western country adopts as its premise the idea of a national spatial system, or set of systems, as the basis for development planning and policy-making; and, for the first time, sectoral planning is clearly relegated to a secondary role.

---

[20]Several geographers, including Professor Kiuchi, have been members of the interdisciplinary Japan Center for Area Development Research in Tokyo, and several foreign geographers, David Kornhauser, Rhoads Murphey, and myself, also have participated in programs sponsored by the Japan Center. See, for example, the report on the Center's Third International Symposium on Regional Development (Tokyo, 1970), and, if you can stand it, my comments on regional development policies, pp. 277-282, as well as articles by T. G. McGee, Hamzah bin Sendut, David Kornhauser, and Tatsuo Ito.

The locational strategies incorporated within the plan have four major components. One relates to the further expansion and integration of the existing axial core, which is accepted as given. That core consists of seven large merging metropolitan regions extending from Hokkaidō to northern Kyūshū. A second addresses itself to the further internal integration, as through administrative reorganization and the elaboration of sophisticated transportation networks, of each of the several metropolitan regions. A third concerns the development of enlarged and more highly integrated urban regions elsewhere in the country, based on existing urban regional centers, whereby the distinctions between urban and rural can be further minimized. The fourth locational strategy addresses itself to the identification, development, and further integration into a single national system of advantaged areas of specialized production and resources for the production of paddy, the raising of livestock, the scientific management of water and forest resources, and the preservation of largely vacant areas rich in scenic and recreational value. Through these strategies time-distance will be 'drastically shortened throughout Japan, [and] the social, economic, and cultural gaps between different regions will be reduced.'

The implications of these strategies for the developing countries are inescapable, whatever their circumstances. The chief barrier to their use in most poor countries is less national poverty than it is multi-ethnicism and the priorities that tend to be given, as in Malaysia, to national integration and political viability. Japan has the great advantage of a well-developed and long-standing ' idea of the state '; many developing countries do not. Nonetheless, the moral of the story, with which no doubt even Chairman Mao would agree, is that one can learn from the experiences of others in these matters, even though solutions to problems must be based firmly on the specific circumstances that exist in particular areas and in particular countries.

Nor is the Japanese experience so improbable a model for other Asian countries as one might suppose. As is well-known, the Japanese postwar economic miracle took place without reliance upon the automobile, although its role has been enhanced greatly within the past decade; and much of Japan's internal connectivity continues to rely upon cabotage. In 1969, for example, 42.2 per cent of all freight ton-kilometers were accounted for by coastal shipping, an increase in fact from 35.5 per pent in 1955. What kind of a developed country is <u>that</u>? Surely, one different from most other developed countries, but not necessarily all that different from its poorer Asian counterparts, particularly those with archipelagic configurations like Indonesia and the Philippines.[21]

I have referred to these matters in some detail because they help illustrate more than is possible in other ways what seems to be one of the major justifications for exposing ourselves, as scholars and men, to those kinds of ideas and percepts which cannot help but result from immersion within a regional or national cultural context and looking at the world from within it. Without doing so, it is unlikely that we will be able to break our culture bounds, understand the major forces at work in the world from a cross-cultural and

---

[21] Ginsburg, "From Colonialism to National Development . . . ," pp. 18-19.

ultimately global standpoint, and restore to our discipline so much of the lustre it has lost in recent years. In this way, too, one may be able to provide perspectives that may help the Japanese (and others) see themselves as others see them.

And what might that view be?

As a result of what we and they now know, Japan appears to be a medium-sized (by European standards), densely populated, comparatively rich country, which by virtue of its long history and high culture constitutes a major cosmos in its own right. Possessed of an extraordinarily high degree of cultural integrity, it also is characterized nonetheless by a high degree of areal differentiation, a marked structural and spatial dualism, and a tradition of regionalism, which have a major impact on the Japanese perception of their environment and on their assessment of priorities for development and progress. The environments they are creating, whether urban or rural, are bound to differ from their Western counterparts, and they will remain characteristically Japanese. Moreover, their perception of their relations with other parts of the world evolves from their long tradition and from their rapidly changing relations in the economic sphere with most of them. That perception also leads them to place an exceptionally high premium on their relations with Asian countries, especially those of East and Southeast Asia. Possessed of remarkably efficient systems of transportation and communication, both under improvement, and of very high rates of growth and therefore of concomitant propensities toward socioeconomic transformation, the Japanese are confronted with many of the problems that also face the other highly developed countries; but these are definable and soluble only within the context of the Japanese cosmos, the roots of which lie outside of Western Civilization.

For all of these reasons, Japan is highly differentiated from all other areas. There is no other place like it.

The late Charles Abrams, a long-time friend and colleague on several United Nations missions to Japan on urban and regional development problems, would not have, I'm sure, minded my closing with an anecdote about *his* perception of the Japanese, as a stranger to their country on the first of those missions in 1960.

We were standing one evening on the verandah of an inn in Kyōtō, overhanging the waters of the Kamo River. The September night was soft, the moon full and high, and the waters of the stream rippled and burbled over the rounded stones and boulders of its shallow bed. Across the river, silhouetted starkly against the deep purple, cloud-strewn sky, loomed a massive multi-storied structure. On top of it was a huge neon sign in modernized Japanese characters and syllabary, glowing and rhythmically blinking out an incomprehensible (to him) message, brokenly mirrored in the swift-flowing waters.

"Beautiful!" said dear Charley. "Look at how well the Japanese deal with the environment! You don't see signs like <u>that</u> in New York."

"But, Charley," I protested, "if you could read it, you wouldn't say that. After all, it's just a brewery, and the sign simply says: 'DRINK ASAHI BEER!'"

SUMMARY OF THE CONFERENCE

NOVEMBER 11, 1972, MORNING SESSION: DISCUSSION GENERATED BY THE PAPERS OF HARRIS AND PARSONS (MIKESELL PRESIDING)

MIKESELL opened the conference with the general statement on geography and foreign area studies that appears in the preface to this volume. He then called upon Harris for a summary of his paper.

HARRIS indicated that there are at least four respects in which studies of the Soviet Union by geographers differ from other area studies by members of the discipline:

(1) The overwhelming proportion of the serious scholarly literature on this area is by those who live within the area. This situation contrasts sharply with the Caribbean or the Pacific Islands in which almost the entire literature is by scholars from outside these areas, or with the Middle East, Africa, Latin America, and India, in which there are varying proportions of internal and external scientific investigations but in which in general up to the present the more important serious geographic studies have been made predominantly by geographers from outside.

(2) In the geographic literature on the Soviet Union there is a virtual absence of cultural geography. This stands in sharp contrast to the work on all other major world areas. In the development of Marxist geography in the Soviet Union after the October Revolution of 1917, the field of geography was divided into two parts: physical geography, in which laws were those of the physical sciences, and economic geography, in which the concepts of the development of society as defined by Marx and Lenin were predominant.

(3) Detailed field studies by geographers from outside the area are almost unknown, and in this respect the Soviet Union is comparable to China during recent decades. Also, foreign work on the Soviet Union is based primarily on the Soviet literature, on statistical and other sources, and only to a negligible extent on direct field observation. Although many of the foreign geographers who have written on the Soviet Union have visited the country on short trips, they have produced no monographic studies of specific areas based on sustained field observation, mapping, interviewing, and recording of data comparable to the

extensive corpus of such studies for other parts of the world.

(4) There is a substantial body of geographic publications in English translated from works produced in the area in Russian by Soviet authors. The journal Soviet Geography: Review and Translation, in ten issues each year since 1960, provides a substantial cross-section of selected Soviet articles from the geographical periodicals in the Soviet Union.

MIKESELL initiated the discussion of Harris' paper by suggesting that "deceptive insights of grand generalizations" have also impeded understanding in other areas. He cited the misleading assumption about pastoral nomadism in studies of the Middle East as one such example, especially the notion that this mode of livelihood is inherently militant and destructive and that its practitioners have been dominant in Middle Eastern history.

HARRIS added that the once widely held notion that pastoral livelihood preceded farming or was a necessary prerequisite to farming could also be regarded as a misleading generalization.

ENGLISH concurred in the comments about the prominence of pastoral nomads and their presumed destructiveness but added that recent archeological work in the Middle East has thrown doubt on the temporal sequence of farming and pastoral life.

PARSONS suggested that sweeping statements about the Amazon region, either that it is a land of exuberant productivity or that natural conditions there excessively limit development, could be cited as another example of "the deceptive insights of grand generalizations." The Amazon is probably potentially more productive than was suspected a generation ago and more fragile than was thought a decade ago.

MIKESELL offered an additional example by pointing to the theory of K. A. Wittfogel [Oriental Despotism (New Haven: Yale University Press, 1957)] that the difference between "Oriental" and "Occidental" civilization can be explained by the importance of large-scale hydraulic works in the former realm and the consequent involvement of governments in public works that are vulnerable to a wide range of natural hazards.

MURPHEY agreed that Wittfogel's theory could be regarded as a misleading and certainly grand generalization, but added that it had stimulated research on the role of irrigation and water management in several areas and hence could be regarded as having had a positive effect on scholarship, even if most of the work thus inspired had entailed rejection of the Wittfogel thesis.

GINSBURG remarked at this point that he felt that Harris had been unfair to Mackinder in quoting from his book of 1919 out of context. He felt that Mackinder's concepts were more subtle and sophisticated than might be deduced from that particular quotation.

HARRIS replied that the quotation from the 1919 book reflected the concern of experiences during World War I and that Mackinder was perhaps writing out of the anguish of wartime and concern for postwar political arrangements. Yet his work was not as much influenced by wartime passion as Alfred Hettner's study of the British Empire during the same war or Derwent Whittlesey's study of German "geopolitics" during the Second World War. Harris expressed the belief that one might identify two categories of generalization, one presented as a hypothesis, and the other as a fact. Although the first can be stimulating to scholarship, the latter may be an impediment.

SOPHER asked whether Harris could define or describe the impact of Marxist thought on Soviet geography.

HARRIS replied that the impact of Marxism could be described under several headings. First, there is an emphasis on applied geography, on scholarship in the service of society. This encourages a heavy stress on implementation of policies and directives of the Party. Some writers have suggested that an economic determinism has tended to replace the environmental determinism that is much criticized in the Soviet literature. In addition, Soviet work in location theory differs somewhat from American practice, for under a centralized planned economy directed and owned by the state there exists the theoretical possibility of planning industrial locations on a comprehensive national scale on a long-range basis with account taken of all sectors of the economy, and all costs including both external and internal economies and immediate or long-range costs and benefits. However, in practice, industrial location decisions often result in expansion of existing facilities as the most economical means of quickly achieving increased production with minimum outlay of new capital and construction. Finally, the impact of ideology on landscapes shows clearly in urban areas where the relative modesty of central business districts reflects a society that is production-oriented rather than consumption-oriented, and in rural areas where state and collective farms with gigantic fields and large settlements reflect the Soviet preference for the large village or, even better, an agricultural city (*agrogorod*), for the political ideal is an industrial proletariat, even, if possible, in reference to agricultural operations.

MIKESELL expressed interest in Harris' statement on the role of scholars from within and without an area. He indicated that he had been impressed by

André Siegfried's "discovery" of cultural tension in the United States [Les Etats-Unis d'aujourd'hui (Paris, 1924)] at a time when most Americans still expressed faith in the "melting pot" analogy. The outsider might enjoy advantages as well as disadvantages, since the implicit comparative framework of his observations could result in a sharper focus than indigenous scholars would be likely to achieve. He suggested that the collection of accounts of the United States by European visitors, recently edited by Oscar Handlin [This Was America (New York: Harper and Row, 1964)], reveals numerous examples of such insights. We owe much to foreign observers, from the time of J. H. St. John de Crèvecoeur and Alexis de Tocqueville, as MURPHEY suggested, to Gunnar Myrdal whose work [An American Dilemma (New York: Harper and Row, 1944)] was named by SOJA as one of the most impressive recent examples of foreign wisdom about American society.

SOJA extended the discussion to the African context by indicating that nearly all of the research work there has until recently been done by outsiders. He pointed out, however, that most African countries are now giving serious thought to the conditions of access for foreign scholars and the latter are, in turn, being forced to be increasingly concerned about their responsibilities to the areas in which they have worked or wish to work. Some countries now stress that access depends upon a scholar being willing to do work of direct benefit to the area, or being willing to undertake studies defined by national goals and aspirations (even to the extent of accepting specific assignments made by government agencies).

HARRIS added that many countries now make a condition that the foreign scholar must utilize and train residents of the area of his research, in order that an infrastructure of scholarly personnel will be built up within the area. The administrators of developing countries come together on various occasions and share ideas on defining the desired relationships of developing and developed countries and in setting conditions of access.

HARRIS also pointed out that in some fields the restriction of materials to national scholars is fairly explicit. Some archival material may not be made available to foreign scholars until domestic scholars have had an opportunity to examine it and use it in their own publications. Also, in the field of archeology there is now a widespread feeling in some countries that sites should be reserved for exploitation by the scholars of the area itself, and a very general requirement that artifacts excavated by foreign archeologists not be exported.

SOPHER asked about the possibilities of doing research in cultural or historical geography in the USSR, since these fields have not been cultivated by Soviet scholars; and SOJA raised the same question in regard to political geography.

HARRIS replied that cultural geography is covered in part by Soviet ethnographers or other scholars, that there is relatively little interest in historical geography in the USSR (as distinct from history of geography or geographical exploration), perhaps because of concentration of attention on achievements since the Revolution rather than an evaluation of pre-revolutionary Russia. Political geography is little taught and then mainly in reference to the character of particular countries, and hence is not comparable to systematic political geography developed in Western countries. On the other hand, he thought that research on internal administrative-territorial organization of the Soviet Union might be possible for foreign scholars.

SOJA asked about the character of Soviet research on other countries--whether it included cultural or political geography.

HARRIS thought that the intellectual tradition of Soviet geography, with its emphasis on economic and physical geography, does not point in these directions. On the other hand, subjects identified as cultural geography in the United States have been studied by Soviet scholars trained in other disciplines.

In response to SOJA's question about whether the situation in Eastern European countries differed in regard to openness to research by outsiders, HARRIS indicated that a wider range of research topics had been investigated by foreigners in the People's Democracies, most notably in Yugoslavia and Poland (where Western research is well known) but some other socialist countries probably do not differ notably from the Soviet Union in terms of research prospects for foreigners.

MIKESELL opened the next phase of the conference by asking Parsons to make a brief statement on his paper.

PARSONS began by pointing to the relative accessibility of Latin America and the fact that field work by American and European scholars can still be undertaken there without serious restriction or excessive expense--a circumstance that compares favorably with most other areas. In addition, he felt that Latin America had been attractive to Americans and Europeans because it provides an occasion for study of tropical environments, and that in this respect the tremendous diversity and fragility of the Latin American environment, combined with its accessibility, offers an unrivaled opportunity for geographic research. The fact that the Spanish and Portuguese languages, and especially the former, are relatively easy to learn, constitutes an additional comparative advantage, although Parsons added that effective command of one or both of these languages is essential for field work, since English cannot be employed as readily as in India, Africa, or Southeastern Asia.

PARSONS indicated that his perspective is that of a cultural geographer interested in the physical and biological environments and that he does not like to think of geography as being exclusively a social science. He felt that one of the major contributions geographers can make to area studies is their ability to treat geographic factors which often are ignored by social scientists and historians and are of crucial importance in the survival of civilizations. If viewed in this light, Parsons felt that Latin America can be regarded as "an important laboratory, especially because of the explosion of population and rapid development of technology that is evident everywhere." He mentioned the destruction of the Amazon rainforests and the hill soils near Oaxaca as examples of the environmental deterioration that is strikingly evident to someone with geographic training. Parsons added that environmental deterioration might be more advanced here than in the tropical regions of the Old World because of Latin America's more advanced level of modernization and technological development. The fact that one can find here so many chain saws capable of clearing several acres of rainforest per day seems to set Latin America apart from the more conservative tropical areas, where subsistence economies are still prevalent.

PARSONS concluded by pointing out that he had tried to confine his review to the work of North American geographers, but felt that much could also be said of the contribution of European geographers, especially French and German, and now increasingly British too, to Latin American studies. In respect to the quantity of published work on Latin America, he thought that more had been written on this area than any of the regions considered in the conference and that perhaps ten per cent of the dissertations completed in American departments of geography during the past few decades had been devoted to Latin America. On the other hand, since most German and French (and very few American) dissertations are published, the balance among German, French, and English publications has probably been about equal during this period.

MIKESELL began the discussion of Parsons' paper by remarking that Latin Americanists form the largest group of area specialists among American geographers and that they are unique also in having their own organization (Conference of Latin Americanist Geographers). He also suggested that Parsons' paper probably comes close to providing a complete inventory, in that it identifies research concentrations and gaps. He then asked if the "personality" of Latin American research--an emphasis on culture-historical and ecological topics and relative neglect of urban and political problems--could be explained by the character of Latin America or is merely an accident of individual or institutional leadership.

PARSONS replied that he thought both influences had been important--the nature of the Latin American "reality" and the "accident" of the impetus given to Latin American research by Carl Sauer.

LOWENTHAL wondered if this interpretation meant that people went to Latin America not only because of its accessibility but also because of its attraction for the particular kinds of research that Parsons had stressed in his review.

PARSONS thought that this generalization might be true, especially in reference to geographers interested in tropical environments. He added that the historical point of view adopted by most Latin American geographers was an advantage in pursuing this interest, since this perspective or dimension is often neglected by natural scientists, even though the latter usually are better informed on the character of particular environmental features. He felt that the ability and willingness of geographers to examine long-term processes is perhaps their most important asset.

MIKESELL wondered if the "starkness" of cultural change in Latin America--the fact that the indigenous civilizations had been overwhelmed by Hispanic civilization--had channeled research into an historical direction and almost forced investigators to adopt a 16th-century point of departure.

SOPHER suggested that the generalization implicit in this question could not be applied very well to India, in that the Indian geography that one might expect to find, given the historical depth of Indian civilization and the complexity of the Indian social environment, is difficult to find. He felt that a "thin, shallow view of man and land," rather than a more profound view of social and historical circumstances was characteristic of most geographic writing on India.

SOJA asked if research undertaken by Latin Americans displayed the bias that seems to be evident in the work of North American geographers interested in this part of the world.

PARSONS replied that Latin American geographers had probably been most influenced by European geographers, especially French and Germans; yet he also felt that the European traditions adopted by Latin Americans did not differ notably from the orientation of North American scholars concerned with their area.

BROOKFIELD remarked that French geographic research seems to have displayed a shift toward urban problems during the past decade, and HARRIS added that Brazilian geographers had shown a substantial interest in urban problems in recent years.

PARSONS agreed and remarked that he had recently encountered four geographers from Paris working on urban problems in Colombia, and added that shifts of topical emphasis had been paralleled by shifts of areal emphasis. For example, several programs of geographic study are underway in Guatemala whereas virtually nothing is being done in Honduras.

GINSBURG called attention to the recommendation in Parsons' paper that more attention be given to "the evaluation and inventory of land resources." He indicated that his recent address as Past-President of the AAG ["From Colonialism to National Development: Geographic Perspectives on Patterns and Policies," Annals of the Association of American Geographers, Vol. 63 (1973), pp. 1-21] included a suggestion that this issue had been neglected by geographers. He wondered what arguments could be advanced in support of the proposition that land inventories ought to be a major enterprise. He pointed to Parsons' suggestion that "nothing less than a systems analysis of the whole complex of natural and social process is likely to yield background for decision making," and wondered if this recommendation was practical, given the obvious difficulty of conducting research within such a vast and complicated framework, or whether it might be regarded as an impossible charge and undesirable diversion.

PARSONS replied that inventory per se was a mechanical effort, an engineering job, and that he personally would have little interest in such work except in its "design stage." On the other hand, he felt that many countries (Guatemala especially) had been willing to employ geographers in inventory work and hence that it might provide a means for getting young geographers into the field; but he expressed no enthusiasm for any enterprise that would restrict geographers to routine work designed to provide raw material for the analysis of another group of scholars.

GINSBURG again referred to the discussion of this issue in his recent address and quoted the following passage: "Such studies need not be encyclopedic listings, although that is what many of them are. What they ought to be are comprehensive appraisals of natural ecosystems as these can be penetrated, exploited, and managed capably by men." He felt that this is not an engineering problem.

PARSONS replied that such efforts for better or worse were in fact usually undertaken by engineers, and that he was skeptical about the practical value of land inventories undertaken by geographers such as the well-known Puerto Rican Survey [The Rural Land Classification Program of Puerto Rico, Northwestern Studies in Geography, No. 1 (1952)] .

MIKESELL indicated that he was impressed with the comprehensive character of Parsons' review, which provides a clear assessment of the strengths and weaknesses of Latin American research, and wondered if the endorsement of individualism at the end of the paper was in conflict with its prior exposure of neglected issues and both tropical and regional gaps.

PARSONS replied that the statement at the end of his paper was a personal preference, that he thought research was best conducted in an atmosphere of individual freedom and that he was not at all inclined to try to influence the preferences of other scholars.

ENGLISH referred again to the issue of the relative neglect of urban problems by geographers devoted to Latin America, and suggested that the review by Morse, cited by Parsons in his paper, indicates that Latin Americanists trained in other disciplines are making good use of the theory and methods of urban geography.

MIKESELL indicated that the work of Latin American geographers was very useful to cultural and historical geographers, but doubtlessly was less rewarding for persons interested in the issues discussed by Soja.

SOJA accepted this generalization but added that he had derived great profit from John Friedmann's Regional Development Policy: The Case of Venezuela (Cambridge: MIT Press, 1966). Hence he felt that the "gaps" evident in geographic research often are filled by scholars identified with other disciplines. He thought that the key issue is not the strengths or weaknesses of geographic research in particular areas, but whether the framework of area studies provides an appropriate base for comparative studies. He expressed skepticism on this point, mainly because of the presumption of areal specialists that a vast knowledge of the language and history of their areas is a necessary prerequisite for most forms of inquiry.

ENGLISH suggested that the most important barrier to cross-cultural comparative study is not language per se but the lack of reliable quantitative data bases. But he did not want to minimize the importance of language training. "If you do not know the language, and do not have a data base, you are not only deaf, you are also blind."

BROOKFIELD commented that numerous examples could be cited from work on New Guinea and the Pacific realm, where investigators without an existing "data base" had solved the problem by collecting data in the field.

SOJA added that the data base of Africa was not such that one would have expected a major development there of theoretical studies and model building; yet such had indeed happened in recent years.

GINSBURG argued that the research orientation suggested by Soja as being characteristic of African studies was equally well developed in reference to South and Southeastern Asia, and Japan. He added that most scholars working in foreign areas inevitably carry concepts developed in their homelands and hence tend to think comparatively. He felt that most work, or at least the work of high quality, undertaken within an area studies framework is essentially comparative.

SOJA expressed doubt about the validity of this generalization.

MURPHEY thought that the difference between geographical research undertaken in Latin America and in Africa might be explained by the fact that the former area had been a major focus of American attention for several decades whereas the latter had been studied intensively only since World War II and hence reflected the research fashions of a more recent era.

MIKESELL expressed the belief that the standard set by areal specialists, especially in regard to language competence, has served to inhibit comparative study, since such specialists often are unable or unwilling to relate their generalizations to other parts of the world and are seldom well disposed toward the work of "intruders." He thought that a person well informed on a particular process or relationship could conduct research of high value in more than one culture area, and that such efforts should be encouraged and welcomed by area specialists.

SOJA agreed with this statement, and also suggested that much could be said of a policy that encouraged review of the works of area specialists by persons with corresponding topical rather than regional skills, i.e., that an urban geographer concerned with Africa might be a more effective reviewer of a work on Latin American cities than a Latin Americanist who was not well informed on the theories and generalizations of urban geography.

ENGLISH accepted this proposition, but warned against the hazards of "transferred wisdom." He pointed out that Iran now had a gazetteer with 55,000 villages and 93 variables but that a person familiar only with Western categories of census taking and classification could not use this material effectively. He cited Janet Abu-Lughod's Cairo: 1001 Years of the City Victorious (Princeton:

Princeton University Press, 1971) as an illustration of the difficulties of applying Western concepts of social area analysis to a non-Western culture.

SOJA felt that the value of this particular study is not its rejection of Western models but rather that it poses a challenge for the refinement or modification of such models. ENGLISH agreed but felt that the author might have offered a different set of findings if she had not had the benefit of a deep understanding of the particular culture.

WHEATLEY expressed the belief that prolonged study of foreign languages is not a wise investment of time for most social scientists, and that speaking fluency is only an indirect objective of the language training programs of most American universities.

MURPHEY and SOPHER both suggested that the critical constraint in comparative study often is not language per se but rather understanding of the "code of a culture," what supposed universals turn out to mean particularistically.

LOWENTHAL suggested that the problem is one of "disciplinary style," and that scholarly barriers and mystiques may be assertions of virtue and autonomy. Yet, as BROOKFIELD added, certain types of research obviously cannot be conducted without an ability to communicate directly with a people in their language and the skill required in such cases is not merely a disciplinary mystique.

LOWENTHAL replied that a generation ago most geographers would not have felt a need for this kind of communication since they were concerned with material culture rather than behavior.

MIKESELL thought that the issue of area study "versus" comparative study might be clarified by referring to the perspective of systems analysis, which includes attempts to achieve holistic understanding of an entire system and more detailed studies of particular linkages within systems. He thought that the preferences and traditions of area studies probably favored the former approach at the expense of the latter and that both are necessary.

NOVEMBER 11, 1972, AFTERNOON SESSION: DISCUSSION GENERATED BY THE PAPERS OF LOWENTHAL AND BROOKFIELD (MIKESELL PRESIDING)

MIKESELL opened the next session of the Conference by asking Lowenthal to comment on his paper.

LOWENTHAL suggested initially that the Caribbean region might be regarded as a microcosm of Europe; and, since the region is so thoroughly Europeanized, he added that the problem of language study discussed earlier is not a serious handicap. He also pointed out that it is difficult to prepare in advance for a program of research in this region and hence that flexibility and an ability to adjust to conditions on the several islands is mandatory. He indicated that sophistication in this regard cannot be acquired in an American or European university and noted also that it is not possible outside of the Caribbean region to acquire speaking skills in Creole or patois. Further, he indicated that the literature on the Caribbean region displays great complexity since authors of different nationalities (American, English, French, Dutch) have emphasized different topics and West Indian scholars have only recently begun to make important contributions. In contrast to the situation described by Harris, he thought that 80-90 per cent of the published work on the Caribbean region had been produced by outsiders. As a consequence of this inbalance, and the pervasive atmosphere of post-colonial nationalism, work by West Indians tends to display a commitment to the presumed needs or aspirations of the individual countries, and yet cannot be wholly independent of American or European influence because of the small size of the national constituencies and the prestige that is assigned to foreign journals and foreign degrees. Hence, although eager to develop their national identities, West Indian intellectuals tend to emulate European or American standards and styles.

Lowenthal added that foreign geographers interested in the West Indies generally adopt an historical and cultural perspective, which means that their preferences have been similar to those described by Parsons as being characteristic of scholars devoted to Latin America. He also thought that West Indians, themselves, feel that they are "living in the past" and continue to be influenced by their colonial heritage, even to the extent of attempting to emulate 18th-century European "life styles."

As a footnote to the remarks on "smallness and intimacy" in his paper, Lowenthal quoted from a notice posted on the police sentry box at the entrance to the Yacht Harbor at St. George's, Grenada:

> GRENADA YACHT SERVICES TO ALL GUARDS:
>
> We have repeatedly and forcefully asked you to keep undesirable or obviously drunken people out of the Yard. However, please use your common sense! The Premier, the Governor, and Members of Cabinet are to be received with all respect and courtesy!

MURPHEY initiated the discussion of Lowenthal's paper by suggesting that his comments on the complexity and connectivity of his area and the relevance to geography of the insights of other disciplines could be applied equally well to the other regions considered during the conference and hence were not peculiar to the West Indies. Yet he also thought it might be significant that these comments were expressed most forcefully by Lowenthal.

LOWENTHAL replied that the generalizations mentioned by Murphey were also presented forcefully in Brookfield's paper, and suggested that complexity and connectivity were perhaps manifest in island realms, and that the small scale of island communities probably heightens awareness of complexity, although he felt that the complexity of social institutions in the Caribbean region probably does not approach the situation in South Asia or the Middle East.

GINSBURG called attention to the comment in his paper that "one gets a sense of wholeness in Japan and in the writings of Japanese geographers that one may not find elsewhere," but added that this impression might be misleading.

LOWENTHAL replied that the internal impression of West Indians is that they are enmeshed in European systems and that outsiders, in contrast, tended to be impressed by the richness of indigenous institutions.

GINSBURG described a recent experience as a participant in a conference at Kingston, Jamaica, on Caribbean regional maritime development. In the course of this conference the West Indian participants were invited to consider an offer from Woods Hole to undertake a complete oceanographic and marine biologic study of the Caribbean, including the use of a modern marine surveying vessel. This offer produced a "state of shock" among the West Indians, since its acceptance would have entailed a degree of cooperation among the several nations and a recognition of regional identity that they had not been able to achieve, and the proposal could also be regarded as an "American intrusion."

LOWENTHAL noted a similar case involving satellite transmission for educational purposes, an offer that was bitterly denounced in one Latin American country as an expression of neo-colonialism.

MIKESELL remarked that the discussion of Parsons' paper had focussed on the "personality" of Latin American research--particularly on its environmental and historical emphasis, and that English's paper had demonstrated an emphasis on the relationship among urban, rural, and pastoral sub-cultures. He wondered whether the Caribbean region might be regarded as an ideal "laboratory" for the study of plural societies and race relations. He indicated that he had found the literature on pluralism in the West Indies, especially M. G. Smith's _The Plural Society in the British West Indies_ (Berkeley: University of California Press, 1965) and Lowenthal's _West Indian Societies_ (London and New York: Oxford University Press, 1972) to be more useful than the more frequently cited works on Southeastern Asia.

LOWENTHAL replied that he could not disagree with the comment about the attractiveness of his area, but felt that this issue was subsumed in his paper under the headings of "scale," "location," and "connection with the outside world." He also felt that the topics mentioned by Mikesell were not peculiarly geographic and had attracted the attention of anthropologists, sociologists, psychologists, and other social scientists.

WHEATLEY expressed surprise at Lowenthal's comment that the requisite skills for Caribbean studies could not be acquired outside the region, since "geographic, sociological, or N-skills" could presumably be acquired prior to arrival in the area.

LOWENTHAL agreed with Wheatley's general proposition, but added that one could not learn Creole or patois in the United States, and that the same could be said for a wide range of problems having to do with living in small, hybrid societies.

GINSBURG suggested that a distinction could be made in this context between areas with a substantial literature and literary tradition (e.g., China) and areas without such literature (e.g., the Philippines).

LOWENTHAL replied that the issue of prior study and preparation for research might better be focused by reference to "high cultures" with a central tradition that is learned and transmitted through a corpus of literature and set of philosophies, and those cultures for which this is not true. He agreed that this would represent a fundamentally different situation for researchers. Yet he felt that in the Caribbean region, as also in the Pacific realm and Latin

America, one is dealing with hybrid or transferred cultures with a substantial European flavor, and that it is difficult to comprehend these situations from a distance.

MIKESELL asked if a person with a good knowledge of French would have difficulty comprehending or adjusting to Haitian Creole or other comparable creolized languages.

LOWENTHAL indicated that such knowledge could be acquired in six to nine months but that the nuances of relationships between Haitians who do and those who do not speak French or Creole would require considerable sophistication about Haitian culture.

PARSONS thought that the skills mentioned by Lowenthal would be acquired incidentally by most people--that they would be "picked up" in the course of field work.

MIKESELL suggested that this attitude would probably not apply to India where an understanding of one of the Indian languages and Hinduism would be accepted by most geographers as a necessary or at least desirable prior commitment.

SOJA remarked that Africa includes both "high" and "primitive" cultures, but he did not feel that investigators have behaved differently in the respective areas.

MIKESELL wondered if work in areas without a heavy weight of indigenous culture tends to encourage a functional rather than an historical perspective.

LOWENTHAL speculated that the distinction being discussed might be more accurately described in reference to the ways in which European or American investigators approach those cultures they regard as being derivates of their own, in contrast to those that are more clearly "indigenous."

MIKESELL asked whether comprehensive study of cultures was possible in a pluralistic situation, since in the Caribbean context this would require intensive study of the antecedent cultures of Indians, Portuguese, British, and even Lebanese.

LOWENTHAL replied that plural societies in the West Indies are not like mosaic societies in the Middle East, since the groups of outside origin adjust rapidly to local situations and no longer offer reliable evidence of their original cultures.

BROOKFIELD reiterated that he did not feel it is possible to speak of a dichotomy between European or non-Europeanized peoples or between primitive and more elaborate cultures, that it would be more profitable to think of a continuum.

LOWENTHAL still felt that areas dominated by values familiar to Europeans and Americans probably have been approached differently from areas that have not been so dominated, and that the Caribbean region clearly belongs in the former category.

MIKESELL then asked if Lowenthal would characterize the West Indies as a region offering exceptional value for studies of race relations, and whether knowledge gained in this area could be used elsewhere.

LOWENTHAL replied that the whole of the Americas, from Brazil to the United States, could be seen in terms of sets of similarities and differences in race relations, and that "the vast range of racial situations in the Caribbean makes it tremendously serviceable as a laboratory for the study of race relations and racial awareness." On the other hand, he doubted that race relations could be described as a systematic body of knowledge.

WHEATLEY agreed, adding that interest in race relations was apt to come as a "spin-off" from the study of roles, or some other aspect of culture.

LOWENTHAL added that racial categories and ethnicity are defined differently in particular contexts and hence cannot be separated from a more general study of cultural contexts.

WHEATLEY added that any category enlarged to include all morphological variants would cease to be a category.

GINSBURG suggested that the implications of this remark are "enormous," since it implies the impossibility of any universal generalization about culture.

MIKESELL expressed dismay at the prospect of a continuum of unique relations that could be examined only in local terms.

LOWENTHAL responded to these several comments by citing the example of Carl N. Degler's comparison of race relations in Brazil and the United States, which required a study of the history of slavery and the connection between race relations and a number of other social categories. Hence his book [Neither Black nor White: Slavery and Race Relations in Brazil and the United States (New York: Macmillan, 1971)] is not just a study of race relations but rather an historically organized social study of the respective countries. He also pointed out that the best comparative study of race relations, Pierre Van den Berghe's Race and Racism (New York: Wiley, 1967) includes consideration of class structure and many other issues that transcend race per se.

SOPHER remarked that the discussion seemed to have taken Lowenthal away from his initial assertion of the relative simplicity of the Caribbean region, and he added that the complexity and subtlety now being recognized is evident in Harald Eidheim's study of a Norwegian community of covert Lappish origin ["When Ethnic Identity Is a Social Stigma," in Fredrick Barth (ed.), Ethnic Groups and Boundries (Boston: Little, Brown, 1969)]. He wondered if the Portuguese in the West Indies also were identified by subtle cultural cues as well as their label.

LOWENTHAL replied that the behavior and cultural cues of the Caribbean "Portuguese" are consequent on their reception by and treatment of the rest of Caribbean society rather than by antecedent traditions in Portugal or, in their case, Madeira.

MIKESELL called attention to Shepard Clough's Basic Values of Western Civilization (New York: Columbia University Press, 1960) in which an attempt is made to categorize this common denominator under several headings: "The End of Man Is Man," "The Glorification of Progress," and so on. He felt that the issue being discussed was that Clough's "basic values" could be taken as a given if one were in a Norwegian village, or in the West Indies, but not if one were in an Arab village or in China.

WHEATLEY expressed the view that the issue of scale in dealing with values forces one to recognize the distinction between science and ethno-science. He thought that attempts to understand the way people look at themselves and explain events that influence them could not lead to scientific explanation in the accepted sense of the term. He thought it unlikely that any universal generalization of scientific validity could be made about culture.

SOPHER objected that science can also be regarded as merely Western "ethno-science."

GINSBURG questioned the generalization that race relations might not be a valid universal topic of investigation since races, however defined, are distributed all over the world, and few societies are homogeneous from the racial point of view.

LOWENTHAL replied that race is a cultural category highly variable in space and time and hence cannot be regarded as a universal.

MIKESELL introduced the next phase of the conference by asking Brookfield to comment on his paper.

BROOKFIELD began by suggesting that the approach taken in his paper was different from some of the others and that much of this difference could be explained by the fact that the Pacific realm is small in population and total land area, possesses considerable cultural variety, has been deeply penetrated by external forces, and shows considerable diversity in reaction to this penetration.

He felt that several points made in his paper could be expanded. In the first place, he thought that a distinctive kind of geography had developed among scholars working on the Pacific Islands, even though a variety of viewpoints and traditions had been brought to the area--Berkeley-style cultural geography, British land-use training, highly quantitative approaches in the last few years, and so on. Nevertheless he felt that the work that has come out of the Pacific area has had a distinctive character. This character includes especially a concern for behavioral response investigated at very small scales. This micro-scale approach, designed to proceed inductively toward comparative study, was adopted promptly by most investigators and continues to be a distinctive feature of Pacific studies.

BROOKFIELD felt that this combination of preferences was both desirable and necessary given the small size of most of the Pacific Islands and the enormous cultural diversity encountered in the larger land areas, especially New Guinea where 700 languages can be identified among the island's three-and-a-half million inhabitants, and pre-colonial social groupings did not exceed a few thousand persons. He felt that studies of man-land relations or spatial organization in an environment of this order require an extremely fine scale of resolution, and that generalization can be achieved in this context only by a carefully controlled process consisting first of the breaking down of culture into its component parts, and then of reaching toward studies of larger scale by comparison.

BROOKFIELD suggested that the second distinctive characteristic of geographers working in the Pacific is a high degree of awareness of the contributions of other scholars devoted to the region and perhaps a corresponding indifference to research undertaken in other parts of the world. He thought that Pacific specialists are inward-looking as a group and the comparative method they have applied within their area has seldom been extended beyond its conventional limits, e.g., to Southeastern Asia or Africa. Brookfield speculated that these preferences or modes of thought--extensive interdisciplinary reading in regard to a particular area combined with indifference to work even within one's own discipline that refers to other areas--has been a common characteristic of foreign area specialists. Hence although the area studies tradition has resolved many of the problems created by disciplinary compartmentalization, it has also encouraged a new set of problems created by areal compartmentalization. Brookfield mentioned "cultural ecology" as a generalizing approach designed to overcome both disciplinary and areal confinement, and expressed the hope that other interdisciplinary and inter-area approaches would be developed by social scientists.

BROOKFIELD's third point of elaboration had to do with the issue of "academic colonialism," or, as a colleague living on Fiji suggested, "academic hijacking"--the use of an area as a laboratory for research that is published elsewhere and is directed to an external constituency. Brookfield suggested that this issue might be addressed best by trying to answer questions about the purpose of area studies. Are such studies justified primarily by their value to the area or merely to the practitioners or students of one's discipline? Or, alternatively, is the audience of area study defined in reference to an ideal of interdisciplinary wisdom, in which case the test of the value of geographic work would be its use by scholars devoted to the same area but trained in other fields? Brookfield thought these questions were important and that the difficulty encountered in answering them had cast doubt on the validity of the area studies tradition. Moreover, he thought that area specialists who have accepted the idea that they should not merely export research--that they should transmit their findings to local authorities and train local students--are not always convinced of the utility or even the validity of their knowledge. Such scholars often face a difficult dilemma, since the methodologies highly regarded by their peers may be difficult to apply in an atmosphere of localism and pragmatism, and in fact are usually designed for the edification of colleagues and students who can only function as spectators. Therefore, Brookfield felt that the issue of academic colonialism relates primarily to the problem of transferring technologies, and he thought that the scholar who wishes to import to his adopted area as well as merely export from it might have great difficulty achieving an acceptable rationale. He cited a recent essay entitled "Dualism Revisited" by Hans Singer in the Journal of Development Studies [Vol. 7 (1970), pp. 60-75], which indicates that approximately 95 per cent of the research and development expenditure of the world is made in the United States and Europe and that the models that determine this effort are based upon high income/high consumption societies, computer-based research, and other circumstances or capabilities that are not widely available in the rest of the world. Although this generalization applies mainly to industrial research and development, Brookfield felt that the problem of transferring academic technologies presents a host of difficulties that should be considered by area specialists, and that the establishment of research centers in the areas themselves should be regarded as a necessary prerequisite for the achievement of a meaningful relationship between the academic tradition of area studies and the needs, aspirations, and realities of the "third world." He concluded by expressing some specific doubts about the validity or utility of central-place theory in this context.

SOJA began the discussion of Brookfield's paper and summary remarks by suggesting that central-place theory and other contributions of theoretical geography should not be treated as "yes" or "no" propositions, subject to total acceptance or rejection, but rather as conceptual frameworks useful in the organization of thought. He thought the kinds of influences that do not produce an isotropic plane inhabited by economic men--colonial exploitation, concentrations of population in a primate city, cultural pluralism, and so on--demand more rather than less effort to test and refine theory. He added that many of Brookfield's opening remarks on the character of the Pacific environment--social organizations of small scale, great physical and cultural diversity--apply equally well to Africa; yet geographers interested in Africa have not adopted the micro-analysis that Brookfield described as being both characteristic of and appropriate to Pacific Studies. He suggested that the strong anthropological tradition of research at this scale may be one of the possible reasons why geographers in Africa have not widely adopted a micro-analytical perspective. Furthermore, the emergence of nation-building as a dominating concept in Africa after World War II attracted many geographers into political and economic studies on a larger scale.

MURPHEY asked if Brookfield felt that people who come from outside an area to study it have an obligation to have their work be of use to the inhabitants of that area.

BROOKFIELD replied that his discussion of this issue was only implicitly programmatic, that he was more interested in the question, seldom posed or answered by areal specialists, of the purpose of their studies.

MURPHEY accepted the validity of the question, but indicated that he would be reluctant to place this burden on all area specialists, since the interests of such scholars might be scientifically valid and yet be of little interest to the inhabitants of an area. He thought that under ideal conditions scholars should have, if not the right and privilege, then at least the ability to study any area for their own purposes without having to justify the results to the people of the area, provided that they do not take advantage of such people or harm them in any way.

PARSONS added that it is easy to forget the fact that most scholars are hedonists, and added in regard to his own motives that he had seldom thought initially of the practical or local value of his research, although this had occasionally happened accidentally.

ENGLISH observed that many scholars interested in international studies have had an unconscious desire to reject the modern world, and have had a romantic attitude toward the conditions of a prior epoch. He thought that this attitude has been evident in the work of many of the cultural geographers trained at Berkeley. Conversely, he felt that scholars more attracted to the practical problems of their own generation had probably devoted themselves to topics that do not require foreign area specialization.

HARRIS pointed out that the Soviet Union is not a suitable area for someone who wants to reject the modern world!

GINSBURG suggested that "academic colonization" has encouraged a counter-trend in many countries, namely "academic nationalism," which calls for a distinctively "Malaysian," "Indonesian," or some other national expression of a particular discipline. He felt that this policy is especially evident in regard to attitudes about training programs.

SOJA indicated that negative attitudes toward "academic colonialism" have been encouraged by a widespread belief that research undertaken by Western scholars has been used by colonial administrators whose interest was to control or exploit local areas. He also felt that a certain paranoia about Western scholars as actual or potential spies has colored thinking throughout the non-Western world.

ENGLISH added that examples of arrogance and insensitivity among Western scholars are easy to cite for the Middle East, including an abortive attempt to excavate a Muslim cemetery!

LOWENTHAL pointed out that Western scholars have, of course, also been engaged in restoration and rescue work in regard to non-Western cultures.

GINSBURG, changing the subject, called attention to a statement in the preliminary version of Brookfield's paper, in which area study was described as having a dual role, that of increasing knowledge of a foreign area, and of constructing a necessary framework for interdisciplinary study. He thought that Brookfield's assertion that these roles were separate and distinctive was too sharply drawn, since efforts at reconciliation had been made and could readily be made.

BROOKFIELD agreed that this distinction should not be so sharply drawn, and that the two objectives were indeed interconnected and complementary.

MIKESELL returned to Brookfield's initial proposition about the distinctiveness of Pacific research, and suggested that the geographers and anthropologists who have worked in the New Guinea Highlands seem to have created a special syndrome. He thought that this syndrome includes a commitment to systems analysis as a conceptual framework, a consequent preoccupation with functions, and a lack of interest in processes of more than very short duration. He also thought that the work done in New Guinea has demonstrated the possibilities of quantitative analysis in an area for which massive numerical data of the type commonly used in the Occident are not available, i.e., that the New Guinea field workers have collected as well as merely used such data. Since geographers and anthropologists with field experience in New Guinea are now fairly numerous and have played an important role in the development of culture ecology as a hybrid discipline, he thought that the strength and weakness of the "New Guinea syndrome" are of major interest.

BROOKFIELD accepted Mikesell's characterization as being basically valid, especially the comment on functionalism, but thought his impression of a lack of historical interest was misleading, since few historical documents are available for the Highlands, and scholars interested in the probable sequence of the development of land use on the island have at least speculated about long-term trends. He also pointed to the historical dimensions in his most recent publication, Colonialism, Development and Independence: The Case of the Melanesian Islands in the South Pacific (Cambridge: University Press, 1972).

MIKESELL indicated that he had thought especially of Roy Rappaport's Pigs for the Ancestors: Ritual in the Ecology of a New Guinea People (New Haven: Yale University Press, 1967), and Eric Waddell's The Mound Builders: Agricultural Practices, Environment and Society in the Central Highlands of New Guinea (Seattle: University of Washington Press, 1972) when he made his generalization about a "New Guinea syndrome."

NOVEMBER 12, 1972, MORNING SESSION: DISCUSSION GENERATED BY THE PAPERS OF MURPHEY, SOPHER, AND ENGLISH (WHEATLEY PRESIDING)

WHEATLEY opened the second day of the conference by suggesting that China might be the most idiosyncratic region under consideration, both in its philosophy and its view of outsiders. He then called upon Murphey for a statement on his paper.

MURPHEY began by agreeing with Wheatley's comment on the idiosyncrasy of China and added that the country was likely to continue to be closed in regard to serious research by outsiders for at least another decade. He indicated that he felt somewhat apologetic about the evangelical and critical tone of his paper but found it difficult to be enthusiastic about the quantity or quality of geographic writing on China. On the other hand, he thought it would be fair to suggest that complaints about the quality of geographic research on China could also be directed to most geographic research undertaken in the 1930's and 1940's.

MURPHEY then turned to the issue of language training discussed earlier in the conference. He pointed out that students interested in China face a difficult problem in that programs designed to produce speaking fluency, which would be desired by someone wishing to undertake field studies in Taiwan, are not widely available in the United States, whereas courses designed to produce competence in reading literary texts are not useful for most categories of social science research. Difficulties produced by the contrary situation--programs designed to produce speaking rather than reading knowledge--have been correspondingly frustrating to other prospective specialists on China.

Nevertheless, he felt that it was unfortunate, given the great contemporary importance of China, the diversity of its environment, and the richness of its civilization, that so few geographers had elected to specialize on this part of the world. He thought that American indifference to China, which had been enhanced by the national policy of "non-recognition," had encouraged an ignorance that is not merely abysmal but "criminal." He felt that much could have been done even during the period of China's isolation by geographers willing to work with published materials, and added that it is difficult to accept as "universal" any geographic theory or concept that has not been applied to one-quarter of humanity.

GINSBURG commented at this point on his own career, and pointed out that he had virtually no constituency as a geographer when he served as Secretary of the Joint Committee on Contemporary China of SSRC and ACLS. He then called attention to the dissertation study of Joseph J. B. Whitney [China: Area, Administration, and Nation Building, University of Chicago, Department of Geography Research Paper No. 123 (1970)], and described it as being exemplary of what geographers can do with Chinese data. He suggested that Whitney's work pointed to "a new direction in political geography" and that his concern with "the organization of area" might be the most appropriate emphasis for a geographer.

Turning to Murphey's contribution, GINSBURG expressed doubt about the suggestion that Communist China had reversed the traditional conception of harmony between man and nature. He thought it was doubtful that the Chinese had really perceived or maintained such a harmony--which he described as a myth--and thought that the issue of what had happened to the Chinese environment had not been studied adequately by geographers. He mentioned Yi-Fu Tuan's recent book on China (Chicago: Aldine, 1969) in this context. SOPHER indicated that Tuan had dealt with the theme of human impact on environment in other publications ["Discrepancies between Environmental Attitude and Behavior: Some Examples from Europe and China," Canadian Geographer, Vol. 12 (1968), pp. 176-191; "Our Treatment of the Environment in Ideal and Actuality," American Scientist, May-June, 1970, pp. 244-249; and Man and Nature, Commission on College Geography, Resource Paper No. 10 (Washington, D.C., 1971)].

MIKESELL expressed the opinion that the reality of environmental modification in China, as distinct from philosophical or perceptual issues, is not a suitable topic for "armchair geographers" and that its neglect is probably a consequence of a lack of opportunity for field investigation.

BROOKFIELD responded to this comment by mentioning the work of Louis Veilleux, who did a Ph.D. dissertation for Hong Kong University on the Chinese paper industry which entailed use of Chinese sources and observations in China. He added that Veilleux has returned to China to investigate the Chinese approach to environmental problems and suggested that his studies represent a new style of North American work in China.

GINSBURG concluded his remarks by suggesting that he was less inclined than Murphey to describe Skinner's work on "Marketing and Social Organization in Rural China" as being exemplary, especially given its regional specificity within China. He also voiced his skepticism about the validity of Skinner's model as a reliable guide to recent or future trends in the organization of area in the developing countries.

HARRIS indicated that several of Murphey's generalizations about China contrast sharply with his impression of the Soviet Union and he found this especially interesting, since one might have supposed, given the common heritage of Marxism, that one would be impressed by similarities. He was impressed that Communist thought in China appeared to differ from that in the Soviet Union in three major respects. First, the present Chinese political leaders seem to emphasize rural life and rural values, whereas the leadership in the Soviet Union appears to emphasize urban development and the urban proletariat, and the urban environment seems to be considered a more appropriate setting for the achievement of Marxist goals. Indeed in the Soviet Union at times the peasantry has been considered a lumpen mass. In any case, urbanization in the Soviet Union has been very rapid and pervasive. Secondly, the two countries have displayed sharply differing attitudes toward location and scale of industry with emphasis in China on very small industrial units dispersed throughout the rural countryside, as symbolized by the so-called backyard iron furnaces, but with emphasis in the Soviet Union initially on large enterprises in large cities, later criticized as "gigantomania." Some of these new large enterprises actually resulted in new large cities, such as Magnitogorsk. Thirdly, attitudes toward the party apparatus appear to differ. One of Lenin's major political contributions was his development of a full-time professional party cadre. His analysis of the failure of the 1905 uprising led him to conclude that such a cadre was necessary. The resulting party apparatus now constitutes an administrative elite in the USSR, which bears heavy responsibility for developing and implementing policy. Harris thought that the writings of Mao Tse-tung in contrast reveal a profound distrust toward bureaucracy or institutionalized leadership, either of the government or the party. In addition to these current communist ideological differences, he felt that account should be taken of profound national differences in society and culture that have evolved over centuries of independent evolution.

SOPHER returned to the issue of ideology and landscape, and suggested that there might be some hazard in accepting the universality of Mircea Eliade's model, that the paradigmatic work of the gods is implemented by man in the landscape, i.e., that a cosmological model is translated into reality [The Sacred and the Profane (New York: Harcourt, Brace & World, 1959)]. He thought that we could also entertain the possibility of an "anti-world" of the ideal, which would be very different from the real world. He suggested as an example that the ideal of harmony represented in Chinese landscape painting probably has little bearing on where people really lived or the reality of Chinese landscape.

WHEATLEY added that the idea of harmony between man and nature is an elitist notion not really relevant to the life of most Chinese, and, in any case, should be related to the traditional, rather than the modern world. The harmony proclaimed in Chinese philosophy was an existential harmony, not something that one would expect to find in the daily life of peasants, who probably had never heard of this concept.

MURPHEY remarked that the issue of myth and reality in regard to the relationship of man and nature is treated at greater length in one of his earlier papers ["Man and Nature in China," Modern Asian Studies, Vol. 1 (1967), pp. 313-333], and added that he had not been able to do justice to it in the paper prepared for the conference. He agreed that the idea of harmony is myth to a degree, an aspect of traditional rhetoric which has been reversed by the present regime. He indicated that the current ideology entails the assertion that Old China tended to submit to nature and accept reality, whereas New China must attack nature and must not accept a reality that is not conducive to welfare. The assumption implicit in this assertion is that an organized communistic society has greater power than a disorganized, exploited society.

MURPHEY thought that the Leninist and Maoist positions are similar on this particular issue, although he agreed that the traditional Chinese concept of harmony probably cannot be found in Russian philosophy.

MIKESELL pointed to the persistence in Western civilization of the conception of a "Golden Age" or lost state of innocence, and wondered if this nostalgic view would be possible in a revolutionary society.

MURPHEY and HARRIS agreed that nostalgia about the past is clearly not a feature of Maoist and Leninist philosophy, but BROOKFIELD pointed out that Marx himself displayed nostalgia about the pre-capitalist era of mankind.

MIKESELL remarked that some Americans working in China prior to the revolution, e.g., Walter Lowdermilk, had been deeply impressed by the evidence of deforestation and erosion in China, and wondered if this concern is still evident.

MURPHEY replied that many Chinese scholars have retained this concern, but he thought that foreigners in the past had been more inclined to speak of "order" and "careful husbandry" when writing about the Chinese landscape.

HARRIS referred to Murphey's suggestion that agricultural production may have doubled or trebled since 1949, and wondered if there was any solid evidence of this magnitude of increase.

MURPHEY replied that one might assert that there is "solid evidence" for this claim in the physical appearance of the people, but that "harder" or more conventional evidence cannot be offered.

WHEATLEY asked if Murphey had gained clear impressions of changes in the economic system of China during his recent visit.

MURPHEY replied that he had gained a clear impression in May of 1972 of improvement in standard of living--the people seemed to be well fed and well clothed, and that one could also find evidence of the growing availability of transistor radios, wrist watches, bicycles, cameras, and other consumer goods at prices just within the reach of wage earners. While he could not characterize this situation as one of affluence, he felt that China had reached a level of economic security during the past decade. He thought that the people he interviewed were perhaps unaware of the possible hazards of "consumerism"--that increase in production would continue to be desired regardless of its influence on the austere, self-denying principles of the Revolution. On the other hand, he thought that contemporary location theory might continue to be difficult to apply in China because of the diffuse, rural-based character of economic development and national planning.

He also suggested that Chinese planning did not seem to be conducted within a rigorously defined "profit-loss" framework, which means that the economic benefits of centralization and specialization might continue to be denied or disregarded. He thought that efficiency was less important to Chinese planners than creating conditions that would bring as many people as possible into the production process.

HARRIS remarked that the Chinese policy of industrial development contrasted sharply with Soviet policy and practice. The initial industrial development in the Soviet Union was achieved under strong centralized planning with massive capital for urban industrial development accumulated through a heavy "tax" on the rural segment of the population through obligatory deliveries of grain and other agricultural products by farms to the state at low set prices.

MURPHEY agreed with this generalization and added that acute food shortages had apparently not occurred in China during the past decade.

GINSBURG suggested that the Chinese probably were well informed on the costs of petroleum and plutonium production and wondered about the extent to which Murphey's overall impression had been influenced by conditions in the particular areas he had been able to visit. He also thought the commune system, which is designed to promote local self-sufficiency, might account for only a small part of China's total production. He wondered, therefore, if the Chinese economy might be described as dual--with different policies for the national "modern" sector and the local "subsistence" sector.

MURPHEY commented that examples of industrial concentration could be cited--truck factories at Shanghai, port related industries, and so on.

MURPHEY then concluded his remarks by indicating that he had not been able to deal adequately in his conference paper with the ideological issue of rural versus urban values, and mentioned a prior essay on "City and Countryside as Ideological Issues: India and China," Comparative Studies in Society and History, Vol. 14 (1972), pp. 250-267.

WHEATLEY initiated the next phase of the conference by asking Sopher to comment on his paper.

SOPHER began by indicating that he had written his paper with a non-geographic audience in mind and that some of its content might not be of interest to the "intellectually endogamous" group gathered at Chicago, "a group devoted to the maintenance of the purity of its caste line." He added that his attempt to describe work according to a conception of "the ecology of the geographic discipline in India" was experimental and perhaps deliberately provocative. In any case, he thought that South Asia is easily the equal of any of the other areas examined during the conference in its idiosyncrasy and in the peculiarity of the behavior of outsiders who have adopted this area as a focus for their research.

Turning to the problem of language training, SOPHER indicated that a speaking knowledge of Hindi could be acquired by diligent students in about six months and added that he would be more favorably inclined toward a course of study emphasizing the structures and histories of Indian languages. He felt that a course of this sort--which apparently is not offered anywhere--would be more appropriate as a foundation for area specialization than one devoted to a particular language, since it would help prospective Indianists to understand cultural groupings and social communication. He added that the prospect of self-training in the field, which Parsons had endorsed for Latin Americanists, was feasible and perhaps desirable in India, since knowledge of local perceptions would be gained more readily in this context than in the detached atmosphere of a university classroom. SOPHER spoke also of misconceptions and depreciatory attitudes that foreigners often carry to India and indicated that his own progress had been retarded by some extent by such "baggage." He also felt that most introductory textbooks in geography convey a misleading impression of the South Asian world.

SOPHER turned next to the issue of the role that geographers interested in foreign areas might play in national planning, and indicated that he was skeptical about this role for several reasons, including personal experience in planning. He thought that the rationale of the sponsor of the conference (the Norman Wait Harris Foundation) was perhaps more appropriate for geographers, since it stressed the importance of a desire to "understand the peoples of other areas." He thought that applied research directed toward planning goals often is deficient in such human understanding, and hence can be dangerous. He felt that the quality of planning does not necessarily determine its acceptance or implementation and that a favorable response to the work of geographers by planning agencies should not, necessarily, impress or direct us. "Knowledge is power" may not mean that "power is knowledge." Hence Sopher thought that Brookfield's invocation of the "virtue of humility" was especially appropriate for geographers.

SOPHER was less inclined to endorse Brookfield's notion that geographers interested in the Pacific region have been unusually inclined to devote themselves to the literature of other disciplines. He felt that open-mindedness and an outward-looking perspective might be characteristic of all or at least most geographers, regardless of their areal interest, and thought that this was a strength rather than a weakness. Sopher pointed to Donald Campbell's model of a "fish-scale" pattern of overlap as the ideal situation for disciplinary arrangement and "clustering with gaps" as the present or real situation ["Ethnocentrism of Disciplines and the Fish-Scale Model of Omniscience," in Muzafer and Carolyn Sherif (eds.), Interdisciplinary Relationships in the Social Sciences (Chicago: Aldine, 1969)]. Sopher felt that the attitude of geographers should be in harmony with Campbell's "ideal situation." He also thought that the fact that geographers may take more from other disciplines than they offer in return, which has often been described as a fault, has in fact been a virtue--a manifestation of flexibility, openness, and imagination.

Since no one present was inclined to dispute Sopher's generalizations about the virtue of geography, discussion of his paper began parenthetically with anecdotal remarks by Murphey and Ginsburg on the circumstances that encouraged G. W. Skinner to recognize the value of central-place theory.

WHEATLEY then began the discussion of Sopher's presentation by asking if the issue of language training deserved further comment or had already been examined adequately.

HARRIS indicated that he felt that any person working in an area where there is a strong national tradition of scholarship had a clear obligation to learn the language of that nation, but added that India presented a different problem, since the corpus of scholarly literature is mainly in English and knowledge of one of the Indian languages would be useful mainly in field studies that require interviewing and the recording of oral tradition.

SOJA referred to Sopher's earlier comment about the importance of understanding the "code" or "cues" of a culture and felt that this knowledge, rather than languages per se should be sought by area specialists. He added that language training could be used as a "qualifying principle" or "status symbol" in a way that would depreciate generalizations that were not based upon linguistic competence.

BROOKFIELD thought that prescription in regard to linguistic competence is virtually impossible to apply, since the range of geographic research includes such a vast array of specific interests, from land-use mapping and industrial location studies (which might be conducted without linguistic skills) to studies of cultural subtleties that demand total fluency.

ENGLISH felt that knowledge of Arabic is essential for most categories of social science research in the Middle East and pointed out also that such knowledge has the effect of "opening" a substantial area of the earth to intimate and sophisticated research. Hence he felt that knowledge of this language is not only useful but offers a considerable personal reward.

HARRIS attempted to summarize the several opinions by pointing to three types of combinations of personal interest and characteristics of an area: (1) ephemeral interest in an area for comparative purposes, (2) a more specific and prolonged interest in an area with a large corpus of scholarly literature, and (3) interest in field study in an area that does not have a substantial native literature or includes numerous dialects that cannot be studied effectively prior to actual arrival in the area. Harris felt that these three combinations present different obligations and opportunities with regard to language study, and that it would be difficult to speak of requirements or standards without trying to recognize this range of personal prospects.

SOJA indicated that he was not concerned about attitudes toward language competence per se but rather the tendency of areal specialists to dismiss generalizations simply by citing specific knowledge of exceptional cases.

WHEATLEY added that it is not possible logically to challenge general knowledge with specific knowledge, and GINSBURG added that the issue is not one of the opposition between such categories of knowledge but whether one can be certain that he in fact possesses knowledge. He thought that much which has been proclaimed to be general in geography is merely conjecture or hypothesis inadequately subjected to verification.

MIKESELL indicated that he thought areal specialists have displayed ambivalence rather than prejudice in regard to the issue mentioned by Soja and Ginsburg. He pointed out that they often complain about the "descriptive" and "inductive" orientation of "traditional" geographic work on their areas and yet also complain that deductive approaches and model-building fail to provide totally adequate explanations. He thought that the "creative tension" evident in the opposition of general and particular knowledge has been a persistent feature of social science.

MIKESELL then asked Sopher if he could comment on the "linguistic stickiness" of India, the extent to which linguistic groupings have acted and are acting as barriers to national integration. He pointed out that he had made good use of Sopher's early essay on "India's Languages and Religions" [*Focus*, Vol. 6, No. 6 (February, 1956)] in his teaching, and wondered if Sopher could recommend other material on this topic.

SOPHER replied that one of his students had investigated the morphology of an Indian language boundary, using census data, and that scholars in sociolinguistics had undertaken some studies; but he did not feel that this problem had received adequate attention. He pointed out that the Indian situation was unusually complex in that one has to be aware of a hierarchy of linguistic regions and different levels of language use, and that this situation encourages the use of lingua francas. He felt that the only general work on the geography of India's languages [Roland Breton, *Géographie humaine des communautés linguistiques en Inde* (Paris, 1969?)] is essentially a gazetteer and does not do justice to the role of language in defining social groups and social distance. Sopher mentioned John Gumperz at Berkeley and Prabod Pandit at Delhi University as having done work on socio-linguistics; yet he was not aware of much specific interest among Indian geographers in the linguistic geography of their country and found this neglect to be ironical.

HARRIS remarked that India impressed him as being a marvelous laboratory for study of the complex relationship between sociolinguistics and cultural and political geography and indicated that he was also surprised that this issue has been neglected. He pointed out that the Soviet Union and India are the two major examples of countries with political subdivisions (republics in the USSR and states in India) based on languages. Apparently this territorial-administrative division grew out of basic political philosophy in the USSR but was accepted grudgingly in India as a result of regional political pressure, at least in the establishment of several recent states, such as Bujarat and Maharashtra in 1960 out of Bombay State, and the division of Punjab (India) into Punjab and Haryana States in 1966.

SOPHER agreed that this situation is especially interesting, and regretted that it seems not to have drawn the attention of geographers in the respective countries. He thought that the character of geographic research conducted by Indians had been determined almost exclusively by the geographic traditions transmitted to India from Britain and the United States and that this process has not encouraged awareness of research problems that reflect the particular character and complexity of Indian cultures.

WHEATLEY wondered if Sopher's endorsement of the idea that geography is a centrifugal discipline meant that he doubted its capacity to generate indigenous theory.

SOPHER felt that the prospect was one of "systems exchange" and that geographers had shown more willingness than most scholars to move across disciplinary boundaries. He did not feel that geography is deficient in indigenous theory.

GINSBURG suggested that each of the social sciences is characterized to some extent by circumstances of both isolation and exposure.

LOWENTHAL thought that geographers might be unusual only in their awareness and preoccupation with this issue, and their willingness to cite and recommend works by scholars trained in other fields. He indicated that he did not feel uncomfortable or apologetic about these characteristics.

MIKESELL pointed out that the articles by anthropologists in the International Encyclopedia of the Social Sciences display an opposite tendency--an almost total indifference to the work of non-anthropologists.

GINSBURG added that the dissertation study of Whitney, which he had mentioned earlier, draws from political science theory as well as geographic theory, and hence could be mentioned as a good example of the advantage of viewing geography as an "open system."

BROOKFIELD suggested that a geographer concerned primarily with spatial patterns would be more inclined to rely on indigenous theory than a geographer with broader, and, especially, ecological interests.

SOJA wondered if it would be fair to suggest that geographers had indigenous theory in regard to patterns but not in regard to processes.

BROOKFIELD and PARSONS both indicated that they could agree with Soja's proposition, but thought that understanding of processes is not essentially a geographic problem and, in any case, is probably more difficult to achieve than the understandings that are more clearly geographic.

WHEATLEY pointed out that David Harvey [Explanation in Geography (London: Edward Arnold, 1969)] also makes a distinction between spatial and temporal theory and regards only the former as providing geographers with an opportunity for the development of indigenous theory.

SOJA thought that this distinction is too sharply drawn since diffusion theory includes both spatial and temporal dimensions.

WHEATLEY indicated that he had raised the issue of indigenous theory in geography and the centrifugal view of the field because of uncertainty in his mind as to whether these conceptions were particularly apt to be identified with areal specialisms. He also wondered if the evolution of geography might lead eventually to a decline in its interdisciplinary orientation.

SOPHER thought geography has been centripetal as much as centrifugal and that the evolution of the discipline would not be likely to change this situation.

LOWENTHAL expressed agreement with Sopher's interpretation and thought that the integrative function of geography is implicit in the character of the discipline and perhaps also in the temperament of its practitioners.

MIKESELL thought the group should display more optimism on this issue, since it might be claimed that geographers are importing scrap iron and exporting Datsuns!

SOPHER concluded his presentation by reiterating his belief that geographers should be more concerned about the cultural meaning of data offered to them in foreign areas and felt that caution in this respect is especially advisable in regard to India.

WHEATLEY launched the next phase of the conference by calling upon English for a statement on his paper.

ENGLISH began by saying that he had tried to give some indication of how geographers have viewed his region. Hence he felt that he had been concerned principally with what has been done rather than what might or should be done. He thought that the systematic organization of the paper was justified, at least as a bibliographic structure, since it would have been more difficult to identify concentrations of research and research gaps if the paper had been organized by countries or areas. On the other hand, English felt that his paper might have a regional bias in that his own work had been directed mainly to Iran. He added that the issue of regional emphasis in any survey of Middle Eastern studies is especially perplexing, since the limits of this realm cannot be defined to anyone's satisfaction, and few scholars have dealt with more than one of its major civilizations--Arab, Persian, and Turkish. Moreover, a distinction can also be made between North Africa and Southwestern Asia in regard to studies of Arab civilization, the former being predominantly a French enterprise and the latter attracting the attention of scholars of several nationalities.

ENGLISH also thought that an element of cohesion can be found in Middle Eastern studies in that most scholars concerned with this realm have recognized the pervasive influence of Islam, have tended to adopt an historical perspective, and have realized that they are concerned with an area of distinctive environmental characteristics. Hence although it is easy to speak of disparity and conflicting traditions in Middle Eastern studies, it is also possible, at least in the geographic literature, to see recognition of the binding influence of Islam, history, and aridity. However, English thought that few generalizations of interest to social scientists had emerged from Middle Eastern studies and suggested that this deficiency might be explained by the size and diversity of the region and also the lack of reliable or accurate statistics in most of the countries. In any case, he did not feel that more than a handful of Middle Eastern specialists have viewed the area as a "laboratory" for the testing of social science theory.

ENGLISH turned next to the three-part division of his paper--city, village, tribe--and remarked that he was less inclined now than previously to think of these units of Middle Eastern culture as being separate or distinctive. He felt that a blurring of his "trilogy" should be recognized and that this could be explained both as a consequence of economic and social developments during recent years and as a product of new orientations in Middle Eastern studies, which include investigations of the connection between village and urban life and the forces that are encouraging abandonment of pastoral livelihood. English added that he would be most inclined, personally, to welcome studies of urban-rural relationships. He added, however, that it is difficult to be optimistic about the prospects for research in the Middle East, since much of the region is now effectively closed to outsiders, especially Americans, and several of the countries not yet closed are imposing severe restrictions on foreign scholars. Hence he felt ambivalent about encouraging students to specialize on this part of the world.

SOJA opened the discussion of English's paper and presentation by expressing surprise that the concept of an "ecological trilogy" was being dismantled so soon after its construction. He then suggested that Middle Eastern scholars did not impress him as being reluctant to think in general terms or to propose theories. He thought that the willingness of Middle Eastern specialists to speak of "the Islamic city," "the conflict between the desert and the sown," "the nature of nomadism," and also English's concept of a "trilogy" suggested a flair for generalization.

ENGLISH agreed that Middle Eastern scholars had been inclined to think in such terms, but felt that dissatisfaction about traditional generalizations is now strikingly evident, especially in regard to "the Islamic city." English felt that established models of urban forms and functions, such as that proposed by G. E. von Grunebaum, were no longer thought to be useful or even interesting. Yet he added that newer conceptions of urban life and institutions in the Middle East were only beginning to appear in the literature.

WHEATLEY remarked parenthetically that the model of the structure of the Muslim town usually associated with von Grunebaum should be attributed to William and George Marçais.

ENGLISH continued his comment on dissatisfaction with traditional concepts by suggesting that the huge literature on the character of villages is redundant and fails, with rare exceptions, to deal with relations among villages or between villages and towns. Hence he thought the Middle Eastern studies have been deficient in developing a spatial perspective.

MIKESELL commented on the fact that English devoted approximately equal attention in his paper to urban and rural studies and wondered if this balance reflected the weight of the respective literatures.

ENGLISH replied that far more work has been done on villages and that interest in cities has developed mainly in the last few years. He thought it would be possible to compile a list of over a hundred monographic works on villages but perhaps only a dozen studies of this character could be recommended to a social scientist interested in cities and even less could be suggested to someone interested in small towns.

MIKESELL commented that if one added the Arab, Turkish, and Persian dimensions to English's topical division of city, village, and tribe, it would be possible to speak of a matrix with nine cells. He thought that the cell representing Arab-village studies would be the most crowded, although major emphasis could also be identified in regard to tribal studies, especially studies of nomadic tribes. If consideration were given to topics falling outside this geographical matrix, Mikesell thought that major concentrations of research on "Arab nationalism" and "the impact of petroleum development" could be identified.

ENGLISH agreed that nationalism and petroleum had attracted considerable attention, especially among political scientists and economists, but added that he had been much influenced by the substantial literature produced by archeologists on urban evolution and the origins of agriculture. He thought that this work was of greater interest to human geographers.

LOWENTHAL indicated that he was struck by the fact that English had stressed the difference of the Middle Eastern urban scene but had not made this point so forcefully in his treatment of village and tribal scenes.

ENGLISH replied that this emphasis was not deliberate, and added that rural-urban migration is producing conditions in many of the larger Middle Eastern cities that can be found elsewhere. Still he felt that the Middle Eastern literature on traditional as opposed to modern cities invites comparison more readily than the literature on rural or tribal life.

MIKESELL asked if English's comment on the growth of large cities meant that it is now possible to speak of urban hierarchies.

ENGLISH replied that settlement hierarchies comparable to those of the Occident are now evident in Iran and Turkey, but that most of the other Middle Eastern countries still seem to be immature in this respect.

MIKESELL indicated that he thought a good case could be made for the advantages of geographic training in exposing ecological problems in the Middle East. He pointed out that many of the countries have only severely limited prospects for agricultural development. He thought that dry cereal farming has expanded to and even beyond its natural limits and that rural development keyed to irrigation also has to be viewed pessimistically. Hence it is important to know where resources for industrial development are located and in general to understand the environmental foundation for the distribution of wealth and poverty. He thought that most social scientists, especially political scientists, lack this understanding and that geographers have a comparative advantage in this respect that they have been reluctant to realize, perhaps because the "new geography" of our time has tended to disregard the earth science tradition of geography.

PARSONS expressed agreement with this statement and reiterated his earlier remark that he did not like to think of geography as being exclusively a social science. He felt that physical geographers and human geographers with understanding of physical geography had much to contribute to area studies, and called attention to Hans Bobek's work as exemplifying this combination of interest in regard to the Middle East.

ENGLISH indicated that he found it easy to accept the validity of these comments and admitted that his paper probably stressed a social science conception of the nature of geography. He recognized that an environmental conception of geography would probably be more comprehensible to scholars trained in other disciplines and indicated that he had found it difficult to get immediate understanding of social-spatial concepts or appreciation of their importance from non-geographers and occasionally even from geography students. He thought that theory devoted to process and ecology should come from the work of geographers in the Middle East and felt such theory would be taken seriously by other scholars.

HARRIS remarked that the emphasis of research on the Middle East differs notably from what Parsons had described for Latin America; yet he thought that it would be possible to describe the two areas as being similar in some respects, including the importance of language and religion as cultural bonds.

BROOKFIELD added that the evident conservatism of the Middle East and its polycentric cultural character might be explained by the lack of dynamism of the Ottoman Empire, as the comparable conservatism and polycentrism of Latin America might be explained by characteristics of the Spanish Empire.

ENGLISH replied that the slow rate of change in the Middle East could also be explained by the authoritarian character of Islamic societies and the fact that Islam includes so many explicit directives in regard to behavior.

WHEATLEY agreed that Islam entails behavioral directives and encourages attitudes that are not likely to foster innovation. On the other hand, he thought that Latin America is far more secular than the Middle East and hence that it is doubtful that the slowness of economic and social development in these two regions can be explained by similar inhibiting institutions.

SOPHER was less inclined to accept a religious explanation of the conservatism of the Middle East and felt that inherited political institutions have been of at least equal importance. He thought that the conception of Islam as being unreceptive to diffusion or innovation is a reflection of Occidental bias.

WHEATLEY still felt that Islam presents problems in regard to social and especially economic development and pointed to the difficulties encountered in attempts to adapt or circumvent Islamic legal codes, especially in regard to the "unearned increment."

MIKESELL asked if the conception of Islam as encouraging fatalism was mythical or exaggerated in English's view.

ENGLISH replied that the peoples he had studied in rural Iran certainly seemed to be fatalistic, particularly in regard to the limitations of their environment and its hazards. He thought that an Iranian peasant who failed to accept a fatalistic view of his life would be extremely frustrated. In other words, in this case fatalism might represent an accurate perception of the "real world."

MIKESELL suggested that the revolutionary character of some of the Middle Eastern countries requires a re-examination of many of the generalizations about change and resistance to change in the Middle East. He thought that Algeria demonstrates especially that a country can reorganize itself radically and yet still profess to uphold traditional Islamic values.

ENGLISH indicated that the situation in Iran differs notably from the Algerian case, since change is being directed from above and probably should be described as evolution rather than revolution, but he agreed that the revolutionary character of Algeria and the progress of secularization in Turkey and Tunisia should be taken into account in any discussion of the capacity of the Middle Eastern countries for radical development.

NOVEMBER 12, 1972, AFTERNOON SESSION: DISCUSSION GENERATED BY THE PAPERS OF SOJA AND GINSBURG, AND CONCLUDING REMARKS (WHEATLEY PRESIDING)

WHEATLEY opened the final session of the conference by asking Soja to comment on his paper.

SOJA began by indicating that he had felt doubtful about the utility of a conference devoted to area studies; yet he did feel enthusiastic about the value of Africa as a focus for studies of spatial organization and regional development, and had stressed these prospects in his paper. He thought that the major challenge for geographers interested in foreign areas is to narrow their focus with the idea of producing work that might be useful in the development of cross-cultural or multi-regional theory, and that a comparative approach is essential to this objective; he defined this approach as an attempt to ask "similar questions about different areas."

SOJA felt that the major failing of the area studies tradition in geography has been the absence of a clearly defined conceptual focus. He indicated that he was particularly interested in work designed to develop generalizations of a comparative nature. He acknowledged that this bias means that he dealt in his paper with only a small part of the geographic literature on Africa, and indicated that he has been preoccupied especially with studies that might enrich understanding or facilitate research design in areas outside of Africa, i.e., exportable studies. Conversely, he indicated that he has also sought to identify work undertaken outside Africa that might be applied in the African context, i.e., importable studies. Soja declared that he had found the geography of modernization to be a useful frame of reference in searching for work of this character, and added that a good case could be made for the value of scholars who are interested in region*s* rather than *a* region. He thought it would be difficult to train geographers to work in this way within the traditional framework of area studies, and hence that it might be necessary to speak of the "inhibiting character" and "obsolescence" of area studies as a framework for research in the social sciences.

WHEATLEY opened the discussion of Soja's paper and summary remarks by suggesting the need for a definition of "area studies."

SOJA replied that he associated this term or tradition with "intensive regional studies" and the infrastructure of departments and institutes that has encouraged them.

MIKESELL wondered if Bryce Wood's article on "Area Studies" in the *International Encyclopedia of the Social Sciences* (Vol. 1, pp. 401-407) might be helpful, and quoted the following passage: "The basic concept of area studies is that the people of a definable geographical sector, acting in their society and their environment, offer an appropriate unit for scholarly attention." He added that the key phrase in this definition is "appropriate unit," and suggested that only about half of the participants in the conference seemed inclined to accept this point of view.

PARSONS remarked that the concept of area studies could also be regarded as merely an administrative convenience or fund-dispensing device.

LOWENTHAL thought that Wood's statement could not be evaluated without some reference to scale--that China or Latin America might be regarded as "appropriate units" whereas smaller areas might not be so regarded; or, conversely, that one might feel more comfortable about smaller than larger areas.

MIKESELL replied that Wood's discussion of the history of the area studies concept suggests that it evolved as an alternative to preoccupation with Western civilization, and that the "units" mentioned were of comparable scale, e.g., East Asia, Africa, Middle East.

In reply to Parsons' comment about area studies being an administrative convenience, SOJA remarked that this framework has influenced behavior and preferences, and HARRIS added that he had become interested in the Soviet Union prior to the development of any organization or administrative structure designed to promote its study.

GINSBURG pointed out that virtually all programmatic statements on area studies, including Wood's, recognize the necessity of disciplinary specialization, and added that he knew of no area program, at least in the social sciences, that is not multi-disciplinary in character. He thought that the notion of an area specialist devoid of disciplinary skill or orientation has no real foundation, that a person of this sort would be a "straw man."

SOJA thought that the critical issue for geographers was the difficulty created by our topical/regional framework for a person interested in comparative regional studies.

BROOKFIELD shifted the focus of the discussion by commenting on some features of Soja's consideration of the geography of modernization or development. He thought that most work on this topic had not yet evolved beyond description and experimentation, and that a number of issues having to do with cumulative causation and the persistence of inequality had not been faced by geographers. The fact that development tends to pile up in core areas poses particularly difficult questions that have not been answered. He also thought that the institutional or structural aspects of development had not been mastered by geographers, and that the discipline has suffered in this regard from an inability to decide whether its emphasis should be on place and phenomena or on people.

SOJA agreed with these comments, particularly the one directed to the persistence of inequality. He indicated that he shared Brookfield's view that changes described as "development" often have increased rather than decreased inequality, and agreed that understanding of the distribution of wealth and poverty is a major challenge for geography. He described his experience as consisting initially of a desire to apply American generalizations about spatial structure to the East African situation. He found this exercise useful as a foundation for understanding of diffusion, migration, and national integration, but added that he was perhaps more concerned at present with institutional issues and processes than with spatial forms per se.

GINSBURG remarked that the idea that development tends to move from higher to lower levels of an urban hierarchy had been pushed too vigorously by some geographers. He thought that this notion could not be presented as a valid generalization and might be dangerous if adopted uncritically by planners.

SOJA agreed that innovations may not move as one might suspect from the character of a settlement hierarchy, but he thought that cases of this sort did not really challenge the validity of the central-place model. He indicated that he regarded central-place theory as the most powerful structural model currently available to geographers and diffusion theory as our most powerful process model. He felt that efforts to combine these two models offer the best prospect for the development of generalizations in the comparative study of regions, and cited Paul Pederson's article on "Innovation Diffusion within and between National Urban Systems [Geographical Analysis, Vol. 2 (1970), pp. 203-254] as an illustration of this prospect.

WHEATLEY turned next to Ginsburg's paper and asked for a statement from him.

GINSBURG began by referring to the paper of Fisher, who had not been able to attend the conference. He thought that Southeast Asia poses the interesting problem of "incipient regionalization" or how a region comes into being, and felt that geographers should be especially concerned about this process. He called attention also to Fisher's discussion of the development of area studies in Britain and the United States, and suggested that it provides a context for consideration of the issues of rationale and relevance that had been debated during the conference.

GINSBURG then commented on the character of four articles that he had distributed as "background papers": (1) "An Introduction to Japan," prepared for a forthcoming volume on Japan to be published by Columbia University Press; (2) "The Resurrection of Japan," prepared for a discussion of "Important New Factors in U.S. Foreign Policy" at the Center for the Study of Democratic Institutions; (3) "The Evolution and Problems of the American City," a lecture given at Keio University, Tokyo, in 1969; and (4) "From Colonialism to National Development: Geographic Perspectives on Patterns and Policies," his address in 1972 as Past-President of the A. A. G. [Annals of the Association of American Geographers, Vol. 63 (1973), pp. 1-21].

GINSBURG remarked that the first paper demonstrates the importance of traditional geographic concepts of location and areal association and that the second could be regarded as a continuation of the traditional interest among geographers in problems of foreign policy. He thought that the third paper could be regarded as an illustration of comparative method, since he had tried in this lecture to explain American cities to a Japanese audience by referring to comparable or contrasting conditions in Japan. He indicated that the fourth work, devoted in large part to development, was distributed in anticipation of the interest in this topic that had indeed been evident during the conference.

GINSBURG then commented on his preferences as a geographer and his experience in foreign areas. He indicated that he had always been interested in "general propositions," particularly in reference to the relations between urban development and policy. He added that he has tended to look at urban and rural institutions as complements rather than separate entities. In regard to development, Ginsburg indicated that he has been interested mainly in processes of economic development and the organization of area. He thought that area study is an appropriate framework for geographic research in that it offers an opportunity to deal with both realities and potentialities, and referred to Whitney's work on China, which had been mentioned earlier in other contexts, as an attractive illustration of an "association of concepts and awareness of reality." He also referred to The City as a Center of Change in Asia, edited by D. J.

Dwyer [Hong Kong: University Press, 1972] as offering numerous examples of attempts to understand "complex systems (broadly social) through application of geographic concepts consisting of general propositions dealing with distributions and ecological relationships."

GINSBURG related that his first foreign area commitment was India, that he then became interested in Japan and China, shifted to Southeastern Asia after the closing of China, and has been moving back and forth between Japan and Southeastern Asia in recent years. He suggested that his perspective in each of these areas has been comparative, and added that he could not think of urban problems in India or Japan without also thinking of such problems elsewhere in Asia and in the United States. He thought that foreign area specialization is of great value in this effort because of the difficulty of transferring concepts developed in the United States or Europe to other areas. On the other hand, he did not feel that area specialization could be practiced without a corresponding commitment to a topical specialization and doubted that comprehensive knowledge of an area should be sought or, if sought, could ever be attained. He felt that work in more than one culture "offers the advantage of hybrid perspective and heightened sensitivity."

GINSBURG added that he felt strongly that geographers should "keep their options open" in regard to research prospects and be skeptical about "common sense wisdom" derived from Western experience. He thought that it is especially dangerous to believe that "cities are the same everywhere, that the same processes operate upon them, and that cultural differences are of no importance." Ginsburg then pointed to several statements in his paper that were designed to provide illustrations of the validity of this principle.

GINSBURG concluded his remarks by indicating that he did not feel that he could make a specific recommendation on the number of Japanese specialists needed in American geography departments but hoped that "there could continue to be at least a minimum-size cadre that will permit us to maintain our point of entry into this important field, without depending on translations of the works of Japanese geographers."

WHEATLEY thanked Ginsburg for his presentation and then attempted to summarize some of the issues considered during the conference. He thought that most of the authors had displayed "moral ambivalence" in regard to working in foreign areas. On the other hand, he felt that the issue of whether one can hope to achieve a thorough understanding of another culture had been treated only indirectly or implicitly by the several authors. Wheatley added that discussion of the intellectual or scientific objectives of foreign area research had produced some disagreement, although the range of opinion expressed on this issue could probably be explained in large part by personal preferences. He concluded by suggesting that the issue most persistently considered during the conference had been the value of foreign area research, and he thought that this assessment had been made under three headings: (1) value to geographic theory, (2) value to the area under consideration, and (3) value to the country of the investigator. He thought that agreement had been reached in regard to the first value, that disagreement had been evident in regard to the second value, and that little consideration had been given to the third value.

MIKESELL indicated that he agreed with Wheatley's summary in general but felt that the issue of the value of foreign area studies to the metropolitan or home country had been considered by Fisher.

LOWENTHAL and MURPHEY added that they felt this value had been taken for granted rather than ignored or denied and that most of the contributors had accepted it implicitly.

WHEATLEY then raised the question of whether a balance could be struck among these three potential or actual values.

GINSBURG replied that he had always assumed that knowledge of foreign areas is of value to the home countries of the people who possess or seek such knowledge and thought that this would be especially true in regard to the United States in view of its extensive pattern of foreign connections and responsibilities. He thought that the reverse of this proposition, that knowledge of foreign areas is not valuable, would be an impossible assertion for a geographer.

MIKESELL suggested that the uneasiness that many scholars now feel about foreign area research is probably a product of the "neo-isolationism" that has evolved as a sequal to "colonialism" and "imperialism," but he added that foreign field study has also been practiced, since the time of Alexander von Humboldt, as a purely academic or scientific enterprise, and he did not believe that this more venerable tradition should give rise to any feelings of guilt. On the other hand, he pointed out that students often assert now that they should not merely export findings--that they should leave something behind, publish in local journals, and, in general, try to be helpful to the people of their adopted area.

HARRIS and MURPHEY pointed out that this attitude displays an element of pretentiousness or "benign arrogance."

BROOKFIELD thought that an obligation to make findings available to the people of a foreign area could be accepted by any scholar, but he felt less comfortable about accepting an obligation to "reform" or "persuade."

WHEATLEY pointed out that publication in London or Chicago might in fact result in more rapid diffusion of "useful information" than publication in a "foreign area."

SOPHER offered the "happy obscurity" of his dissertation, published in Singapore, as evidence in support of Wheatley's proposition.

SOJA indicated that the issue being discussed is no longer a matter of choice in many countries, that contribution to the solution of national problems and full disclosure of findings can be a condition of access.

LOWENTHAL pointed out that a scholar eager to be helpful to the people in the area he is studying might find himself in conflict with established authority.

GINSBURG added that Western scholars interested in policy, and particularly those with evangelical attitudes about their methods or theories, can cause difficulties and even retard development in foreign countries. He thought that this prospect is particularly likely to be realized in areas where deference is displayed toward the wisdom of Western "experts."

WHEATLEY then posed the question of what should be understood by "cross-cultural comparison," which had been mentioned by most of the authors as a desirable objective for scholars interested in foreign areas. He thought that much of what has been called comparison simply represents an attempt to dissolve diversity by generalizing the category of investigation, and he added that the power of an idea should be measured in reference to its ability to order rather than eliminate diversity.

MURPHEY thought that the major objective of comparative study is to compare similar phenomena (e.g., land-holding patterns) in different areas in order to gain a deeper understanding of functions.

LOWENTHAL mentioned the work of Marshall Segall, Donald Campbell, and Melville Herskovitz on The Influence of Culture on Visual Perception (Indianapolis: Bobbs-Merrill, 1966) as an example of controlled comparison made possible by isolating a particular feature and examining its presence or absence in a wide range of environmental and cultural circumstances. He thought that comparison could also be undertaken of far more complicated features or even entire configurations of culture, although not on a statistical basis.

MIKESELL suggested that comparison of particular linkages in ecosystems has been the essential task of cultural ecology and he thought that this task had been accomplished successfully by many geographers.

BROOKFIELD agreed with these remarks on the feasibility and value of controlled comparison. He thought that such comparison permits generalizations about functional relationships in more than one society and suggested that the "sorting-out" process involved in such studies may offer the best prospect for identifying significant research problems in geography.

WHEATLEY expressed skepticism about some of these statements but indicated that he was persuaded that his colleagues were persuaded that comparative study is a desirable and attainable objective for geographers. He then joined with MIKESELL in expressing thanks to the Norman Wait Harris Foundation for its sponsorship of the conference and to the contributors for their willingness to respond to a difficult charge.

THE UNIVERSITY OF CHICAGO
DEPARTMENT OF GEOGRAPHY
RESEARCH PAPERS (Lithographed, 6×9 Inches)

(Available from Department of Geography, The University of Chicago, 5828 S. University Ave., Chicago, Illinois 60637. Price: $5.00 each; by series subscription, $4.00 each.)

48. BOXER, BARUCH. *Israeli Shipping and Foreign Trade* 1957. 176 pp.
53. ACKERMAN, EDWARD A. *Geography as a Fundamental Research Discipline* 1958. 40 pp. $1.00
56. MURPHY, FRANCIS C. *Regulating Flood-Plain Development* 1958. 216 pp.
62. GINSBURG, NORTON, editor. *Essays on Geography and Economic Development* 1960. 196 pp.
71. GILBERT, E. W. *The University Town in England and West Germany* 1961. 79 pp. 4 plates. 30 maps and diagrams. (Free to new subscribers)
72. BOXER, BARUCH. *Ocean Shipping in the Evolution of Hong Kong* 1961. 108 pp.
84. KANSKY, K. J. *Structure of Transportation Networks: Relationships between Network Geometry and Regional Characteristics* 1963. 155 pp.
91. HILL, A. DAVID. *The Changing Landscape of a Mexican Municipio, Villa Las Rosas, Chiapas* NAS-NRC Foreign Field Research Program Report No. 26. 1964. 121 pp.
94. MC MANIS, DOUGLAS R. *The Initial Evaluation and Utilization of the Illinois Prairies, 1815–1840* 1964. 109 pp.
97. BOWDEN, LEONARD W. *Diffusion of the Decision To Irrigate: Simulation of the Spread of a New Resource Management Practice in the Colorado Northern High Plains* 1965. 146 pp.
98. KATES, ROBERT W. *Industrial Flood Losses: Damage Estimation in the Lehigh Valley* 1965. 76 pp.
102. AHMAD, QAZI. *Indian Cities: Characteristics and Correlates* 1965. 184 pp.
103. BARNUM, H. GARDINER. *Market Centers and Hinterlands in Baden-Württemberg* 1966. 172 pp.
105. SEWELL, W. R. DERRICK, *et al. Human Dimensions of Weather Modification* 1966. 423 pp.
106. SAARINEN, THOMAS F. *Perception of the Drought Hazard on the Great Plains* 1966. 183 pp.
107. SOLZMAN, DAVID M. *Waterway Industrial Sites: A Chicago Case Study* 1967. 138 pp.
108. KASPERSON, ROGER E. *The Dodecanese: Diversity and Unity in Island Politics* 1967. 184 pp.
110. REED, WALLACE E. *Areal Interaction in India: Commodity Flows of the Bengal-Bihar Industrial Area* 1967. 210 pp.
112. BOURNE, LARRY S. *Private Redevelopment of the Central City: Spatial Processes of Structural Change in the City of Toronto* 1967. 199 pp.
113. BRUSH, JOHN E., and GAUTHIER, HOWARD L., JR. *Service Centers and Consumer Trips: Studies on the Philadelphia Metropolitan Fringe* 1968. 182 pp.
114. CLARKSON, JAMES D. *The Cultural Ecology of a Chinese Village: Cameron Highlands, Malaysia* 1968. 174 pp.
115. BURTON, IAN, KATES, ROBERT W., and SNEAD, RODMAN E. *The Human Ecology of Coastal Flood Hazard in Megalopolis* 1968. 196 pp.
117. WONG, SHUE TUCK. *Perception of Choice and Factors Affecting Industrial Water Supply Decisions in Northeastern Illinois* 1968. 96 pp.
119. DIENES, LESLIE. *Locational Factors and Locational Developments in the Soviet Chemical Industry* 1969. 285 pp.
120. MIHELIC, DUSAN. *The Political Element in the Port Geography of Trieste* 1969. 104 pp.
121. BAUMANN, DUANE. *The Recreational Use of Domestic Water Supply Reservoirs: Perception and Choice* 1969. 125 pp.
122. LIND, AULIS O. *Coastal Landforms of Cat Island, Bahamas: A Study of Holocene Accretionary Topography and Sea-Level Change* 1969. 156 pp.
123. WHITNEY, JOSEPH. *China: Area, Administration and Nation Building* 1970. 198 pp.
124. EARICKSON, ROBERT. *The Spatial Behavior of Hospital Patients: A Behavioral Approach to Spatial Interaction in Metropolitan Chicago* 1970. 198 pp.
125. DAY, JOHN C. *Managing the Lower Rio Grande: An Experience in International River Development* 1970. 277 pp.
126. MAC IVER, IAN. *Urban Water Supply Alternatives: Perception and Choice in the Grand Basin, Ontario* 1970. 178 pp.

127. GOHEEN, PETER G. *Victorian Toronto, 1850 to 1900: Pattern and Process of Growth* 1970. 278 pp.

128. GOOD, CHARLES M. *Rural Markets and Trade in East Africa* 1970. 252 pp.

129. MEYER, DAVID R. *Spatial Variation of Black Urban Households* 1970. 127 pp.

130. GLADFELTER, BRUCE. *Meseta and Campiña Landforms in Central Spain: A Geomorphology of the Alto Henares Basin* 1971. 204 pp.

131. NEILS, ELAINE M. *Reservation to City: Indian Urbanization and Federal Relocation* 1971. 200 pp.

132. MOLINE, NORMAN T. *Mobility and the Small Town, 1900–1930* 1971. 169 pp.

133. SCHWIND, PAUL J. *Migration and Regional Development in the United States, 1950–1960* 1971. 170 pp.

134. PYLE, GERALD F. *Heart Disease, Cancer and Stroke in Chicago: A Geographical Analysis with Facilities Plans for 1980* 1971. 292 pp.

135. JOHNSON, JAMES F. *Renovated Waste Water: An Alternative Source of Municipal Water Supply in the U.S.* 1971. 155 pp.

136. BUTZER, KARL W. *Recent History of an Ethiopian Delta: The Omo River and the Level of Lake Rudolf* 1971. 184 pp.

137. HARRIS, CHAUNCY D. *Annotated World List of Selected Current Geographical Serials in English, French, and German* 3rd edition 1971. 77 pp.

138. HARRIS, CHAUNCY D., and FELLMANN, JEROME D. *International List of Geographical Serials* 2nd edition 1971. 267 pp.

139. MC MANIS, DOUGLAS R. *European Impressions of the New England Coast, 1497–1620* 1972. 147 pp.

140. COHEN, YEHOSHUA S. *Diffusion of an Innovation in an Urban System: The Spread of Planned Regional Shopping Centers in the United States, 1949–1968* 1972. 136 pp.

141. MITCHELL, NORA. *The Indian Hill-Station: Kodaikanal* 1972. 199 pp.

142. PLATT, RUTHERFORD H. *The Open Space Decision Process: Spatial Allocation of Costs and Benefits* 1972. 189 pp.

143. GOLANT, STEPHEN M. *The Residential Location and Spatial Behavior of the Elderly: A Canadian Example* 1972. 226 pp.

144. PANNELL, CLIFTON W. *T'ai-chung, T'ai-wan: Structure and Function* 1973. 200 pp.

145. LANKFORD, PHILIP M. *Regional Incomes in the United States, 1929–1967: Level, Distribution, Stability, and Growth* 1972. 137 pp.

146. FREEMAN, DONALD B. *International Trade, Migration, and Capital Flows: A Quantitative Analysis of Spatial Economic Interaction* 1973. 202 pp.

147. MYERS, SARAH K. *Language Shift Among Migrants to Lima, Peru* 1973. 204 pp.

148. JOHNSON, DOUGLAS L. *Jabal al-Akhdar, Cyrenaica: An Historical Geography of Settlement and Livelihood* 1973. 240 pp.

149. YEUNG, YUE-MAN. *National Development Policy and Urban Transformation in Singapore: A Study of Public Housing and the Marketing System* 1973. 204 pp.

150. HALL, FRED L. *Location Criteria for High Schools: Student Transportation and Racial Integration* 1973. 156 pp.

151. ROSENBERG, TERRY J. *Residence, Employment, and Mobility of Puerto Ricans in New York City* 1973. 230 pp.

152. MIKESELL, MARVIN W., Editor. *Geographers Abroad: Essays on the Problems and Prospects of Research in Foreign Areas* 1973. 296 pp.